ICIAM2023 Springer Series

Volume 5

This series aims to publish some of the most relevant results presented at the ICIAM 2023 held in Tokyo in August 2023.

The series is managed by an independent Editorial Board, and will include peer-reviewed content only, including the Invited Speakers volume as well as books resulting from mini-symposia and collateral workshops.

The series is aimed at providing useful reference material to academic and researchers at an international level.

Vladimir Simeonov Georgiev • Tohru Ozawa
Editors

Various Methods for the Analysis of PDEs

Springer

Editors
Vladimir Simeonov Georgiev
Department of Mathematics
University of Pisa
Pisa, Italy

Tohru Ozawa
Department of Applied Physics
Waseda University
Shinjuku, Tokyo, Japan

ISSN 3091-3098 ISSN 3091-3101 (electronic)
ICIAM2023 Springer Series
ISBN 978-981-96-9163-0 ISBN 978-981-96-9164-7 (eBook)
https://doi.org/10.1007/978-981-96-9164-7

Mathematics Subject Classification: 26D, 35A, 35J, 35Q, 42B, 49J, 81Q, 35L, 47A, 47N

This Springer imprint is published by the registered company Springer Nature Singapore Pte Ltd.
The registered company address is: 152 Beach Road, #21-01/04 Gateway East, Singapore 189721, Singapore

Preface

We are pleased to present the proceedings of the Mini-symposium "Various Methods in PDEs" at the International Congress on Industrial and Applied Mathematics (ICIAM) 2023, held at Waseda University, Tokyo, Japan. This session brought together a diverse group of researchers and experts to discuss the latest developments and innovations in Partial Differential Equations.

The papers in our Mini-symposium proceedings reflect the high quality and breadth of research presented at the session. They cover a wide range of topics, showcasing the dynamic and interdisciplinary nature of industrial and applied mathematics. Each contribution underwent a rigorous peer-review process to ensure the highest standards of academic excellence.

One of the topics at our Mini-symposium addresses novel contributions to the field of interpolation inequalities, specifically stability results for the Sobolev inequality and the Gaussian logarithmic Sobolev inequality with explicit and dimensionally sharp constants.

Another focus is the study of new Strichartz estimates specifically for orthonormal families of initial data. This is significant because traditional Strichartz estimates typically apply to individual functions. Extending these estimates to orthonormal families provides deeper insights into the behavior of solutions to dispersive equations. The results are not limited to the wave equation but also extend to the Klein-Gordon and fractional Schrödinger equations.

Additionally, the results presented at the Mini-symposium examine the asymptotic behavior of the massive Maxwell-Klein-Gordon system under the Lorenz gauge condition in dimension (1+4). This includes a detailed analysis of the asymptotic behavior of solutions to the mMKG system, along with a scattering result for mMKG.

A new characterization of the time-dependent free Schrödinger operator is also presented. The novelty lies in its unique status as the only linear PDE invariant under the Galilean group in Euclidean space-time.

Furthermore, the proceedings include lifespan estimates for solutions of the damped wave equation. In the case of nonlinearity with a subcritical Fujita exponent, decay estimates for particular initial data are presented.

We hope these proceedings will serve as a valuable resource for researchers and professionals in the field. The insights and findings presented here not only advance our understanding of complex mathematical problems but also highlight the critical role of mathematics in addressing real-world challenges.

We extend our gratitude to all the authors for their contributions, the reviewers for their diligent work, and the participants for their active engagement. We also thank the organizing committee for their dedication and effort in making this Mini-symposium a success.

We look forward to the continued exchange of ideas and collaboration that will drive the future of industrial and applied mathematics.

Pisa, Italy — Vladimir Simeonov Georgiev
Shinjuku, Japan — Tohru Ozawa

Contents

Decay of Solution to 1d Subcritical Damped Wave Equation Under Some Initial Condition

Kazumasa Fujiwara and Vladimir Simeonov Georgiev

Abstract We study decay properties of the solution to the 1d damped wave equation in the Fujita subcritical case $p < 3$. Namely, we consider appropriate small initial data, such that global solution exists even for $p < 3$ and obtain an estimate of the L^∞ norm that gives a decay faster than the classical linear time decay $t^{-1/2}$. We assume initial data in space of type $L^1 \cap L^\infty$ or corresponding Sobolev spaces without imposing some additional space dependent weights in these spaces.

Keywords Damped wave equations · Fujita critical exponent · Power-type nonlinearity · Decay estimate

2020 Mathematics Subject Classification 35B40, 35B33, 35B51

1 Introduction

In this manuscript, we consider the Cauchy problem:

$$\begin{cases} \partial_t^2 u + \partial_t u - \Delta u = |u|^p, & (t, x) \in (0, \infty) \times \mathbb{R}^n, \\ u(0, x) = u_0(x), & x \in \mathbb{R}^n, \\ \partial_t u(0, x) = u_1(x), & x \in \mathbb{R}^n. \end{cases} \tag{1}$$

K. Fujiwara (✉)
Faculty of Advanced Science and Technology, Ryukoku University, Otsu, Shiga, Japan
e-mail: fujiwara.kazumasa@math.ryukoku.ac.jp

V. S. Georgiev
Department of Mathematics, University of Pisa, Pisa, Italy

Faculty of Science and Engineering, Waseda University, Shinjuku-ku, Tokyo, Japan

Institute of Mathematics and Informatics, Bulgarian Academy of Sciences, Sofia, Bulgaria
e-mail: georgiev@dm.unipi.it

V. S. Georgiev, T. Ozawa (eds.), *Various Methods for the Analysis of PDEs*,
ICIAM2023 Springer Series 5, https://doi.org/10.1007/978-981-96-9164-7_1

The aim of this note is to estimate the decay rate of u with $n = 1$ in the L^∞ framework under the pointwise initial condition

$$u_0 \leq 0, \quad u_1 \leq -u_0/2. \tag{2}$$

We remark that the local well-posedness of (1) is known under some initial condition. A classical result can be found in [24].

We firstly note that it is expected that solutions to (1) behaves similarly to those of

$$\begin{cases} \partial_t v - \Delta v = |v|^p, & (t,x) \in (0,\infty) \times \mathbb{R}^n, \\ v(0,x) = u_0(x) + u_1(x), & x \in \mathbb{R}^n \end{cases} \tag{3}$$

from the perspective of perturbation theory. It is because the corresponding linear solution to (1) is asymptotically equivalent to that of

$$\begin{cases} \partial_t v - \Delta v = 0, & (t,x) \in (0,\infty) \times \mathbb{R}^n, \\ v(0,x) = u_0(x) + u_1(x), & x \in \mathbb{R}^n. \end{cases} \tag{4}$$

The asymptotic equivalence was firstly pointed out by Nishihara [22] in the three dimensional case and Marcati and Nishihara [21] in the one dimensional case. Later, Ikeda et al. [14] generalized these results for general dimensional case by using Fourier multiplier method. We note that the local well-posedness of (1) is also obtained by the linear estimates of [14, 21, 22] and standard contraction argument, under a certain initial condition.

We define so-called Fujita exponent $p_F(n)$ by

$$p_F(n) = 1 + \frac{2}{n}.$$

$p_F(n)$ is known to be the critical exponent to the Cauchy problem of Fujita type heat equation: Namely, if $p \leq p_F(n)$, then any non-trivial non-negative valued solutions to (3) blows up in finite time, while if $p > p_F(n)$, some non-negative valued solutions exists globally. For details, see [4, 8, 19]. It has been studied that a similar classification works also for (1).

In one dimensional super critical case, i.e. when $p > p_F(1)$, applying the linear estimates [21, Theorem 1.2], for small initial data, global mild solutions to (1) may be constructed in $C([0,\infty); W^{1,1} \cap W^{1,\infty}) \cap C^1((0,\infty); \times L^1 \cap L^\infty)$ by a standard contraction argument. Here $W^{k,m}$ means the usual Sobolev space defined by

$$W^{k,m} = \{f \in L^m \mid \partial^\alpha f \in L^m \Leftarrow |\alpha| \leq k\},$$

where α is a multi-index. Moreover, it is seen that the solution (1) is asymptotically equivalent to that of (4) with modified v_0 if necessary. For related topics, we refer to [14, 15, 17, 18]. We remark that in higher dimensional case, (1) does not construct a dynamical system in the $W^{1,1} \cap W^{1,\infty}$ framework because of the wave effect. Therefore, in this manuscript, we consider only the one dimensional case for simplicity. We also remark that, before [22], for small compactly supported initial data, Todorova and Yordanov [25] deployed the weighted energy method and showed the global existence in the H^1 framework when

$$p_F(n) < p \begin{cases} \leq \frac{n}{n-2} & \text{if} \quad n \geq 3, \\ < \infty & \text{if} \quad n = 1, 2, \end{cases}$$

where H^s means the Sobolev space defined with Bessel potential by $(1-\Delta)^{-s/2}L^2$.

In the critical and subcritical cases: $1 < p \leq p_F(n)$, it is known that the behavior of solutions to (1) significantly varies not only with the size of the initial values but also with their shape.

In the case where $1 < p \leq p_F(n)$ and $(u_0, u_1) \in L^1 \times L^1$ is regular enough and satisfies

$$\int_{\mathbb{R}^n} (u_0(x) + u_1(x))dx > 0, \tag{5}$$

the corresponding solutions blow up in the L^1 framework. Especially, under the initial condition (5), the maximal existence time T_0 of solutions with $(u_0, u_1) = (\varepsilon f_0, \varepsilon f_1)$ enjoys the estimates

$$c_{n,f_0,f_1} T_{n,p}(\varepsilon) < T_0 < C_{n,f_0,f_1} T_{n,p}(\varepsilon), \tag{6}$$

where

$$T_{n,p}(\varepsilon) = \begin{cases} \varepsilon^{-\frac{2(p-1)}{2-n(p-1)}} & \text{if} \quad p \in (1, p_F(n)), \\ \exp(\varepsilon^{-\frac{2}{n}}) & \text{if} \quad p = p_F(n) \end{cases}$$

and c_{n,f_0,f_1} and C_{n,f_0,f_1} are positive constants depending only on (n, f_0, f_1) but independent of ε. We note that $T_{n,p}(\varepsilon)$ gives the lifespan order of solutions to (3) for non-negative valued initial values as well. We list the related results in the following table:

Li and Zhou [20]	$n \leq 2$	$1 < p \leq p_F(n)$	$T_0 < CT_{n,p}(\varepsilon)$
Nishihara [22]	$n = 3$	$1 < p \leq p_F(n)$	$T_0 < CT_{n,p}(\varepsilon)$
Ikeda and Wakasugi [13]	$n \geq 1$	$1 < p < p_F(n)$	$T_0 < CT_{n,p}(\varepsilon)$
Ikeda and Ogawa [11]	$n \geq 1$	$p = p_F(n)$	$T_0 > CT_{n,p}(\varepsilon)$
Fujiwara et al. [6]	$n \geq 1$	$1 < p < p_F(n)$	$T_0 > CT_{n,p}(\varepsilon)$
Ikeda and Sobajima [12]	$n \geq 1$	$p = p_F(n)$	$T_0 < CT_{n,p}(\varepsilon)$

Therefore, one can say that under the initial condition (5), the lifespan of solutions to (1) is estimated as if $\partial_t^2 u$ is negligible. We remark that the blowup was shown in [20] in the L^∞ framework by using the expansion of the integral kernel of the propagator of damped wave equation $S(t)$, which we see the detail in the next section. We also note that under the initial condition:

$$\int_{\mathbb{R}^n} (u_0(x) + u_1(x))dx = 0, \qquad u_0, u_1 \not\equiv 0, \tag{7}$$

solutions blow up in the L^1 framework as well. In this case, the sharp lifespan estimate was shown in the one dimensional case in [5].

In the case where (u_0, u_1) satisfies neither (5) nor (7), it becomes difficult to grasp the behavior of solutions to (1). We remark that the finite propagation property of (1) implies that for any $\mu \in \mathbb{R}$, there exists rapidly decreasing smooth functions (u_0, u_1) satisfying

$$\int_{\mathbb{R}} (u_0(x) + u_1(x))dx = \mu$$

and the corresponding solution to (1) blows up in finite time. Meanwhile, we remember the following global existence theorem:

Proposition 1.1 ([20, Theorem 1.2], [5, Theorem 1.2]) *Let $p > 1$. Let $u_0 \in W^{1.1} \cap W^{1,p}$ and $u_1 \in L^1 \cap L^p$ satisfy* (2). *There exists $\varepsilon_1 > 0$ such that if*

$$\|u_0\|_{W^{1,1}\cap W^{1,p}} + \|u_1\|_{L^1 \cap L^p} \leq \varepsilon_1,$$

then there exists a non-negative valued mild solution

$$u \in C([0, \infty); W^{1,1} \cap W^{1,p}) \cap C^1((0, \infty); L^1 \cap L^p) \cap L^p((0, \infty) \times \mathbb{R}).$$

to (1). *In addition, u is estimated pointwisely by*

$$\begin{aligned} 0 &\geq \frac{e^{-t/2}}{2}(u_0(x+t) + u_0(x-t)) + \frac{e^{-t/2}}{2} \int_{x-t}^{x+t} u_1(y) + \frac{1}{2}u_0(y)dy \\ &\geq u(t,x) \geq S(t)(u_0 + u_1)(x) + \partial_t S(t) u_0(x) \end{aligned} \tag{8}$$

for any t, x.

Under the initial condition (2), the wave effect guarantees the non-positivity of solutions to (1) and this establishes the global existence. We also note that $u \in L^p((0, \infty) \times \mathbb{R})$ follows from the blowup argument above. Indeed, by integrating Eq. (1) with respect to t, x, we have

$$\int_{\mathbb{R}} u(t,x) + \partial_t u(t,x) dx = \int_{\mathbb{R}} u_0(x) + u_1(x) dx + \int_0^t \int_{\mathbb{R}} |u(\tau, x)|^p dx d\tau.$$

If the nonlinearity is not integrable, then there exists $t_0 > 0$ such that

$$\int_{\mathbb{R}} u(t_0, x) + \partial_t u(t_0, x)dx > 0.$$

Then the blowup argument above implies that the corresponding solution to (1) cannot exists globally. Therefore, the nonlinearity of solutions of Proposition 1.1 is integrable with respect to t and x.

Since solutions u of Proposition 1.1 is pointwisely non-positive, (1) with $n = 1$ is rewritten with $w = -u$ as

$$\begin{cases} \partial_t^2 w + \partial_t w - \Delta w + |w|^{p-1}w = 0, & (t, x) \in (0, \infty) \times \mathbb{R}, \\ w(0, x) = w_0 = -u_0, & x \in \mathbb{R}, \\ \partial_t w(0, x) = w_1 = -u_1, & x \in \mathbb{R}. \end{cases} \tag{9}$$

The behavior of w has been extensively studied even without initial sign condition. Especially it is expected that w is estimated by

$$\|w(t)\|_{L^q} \leq C \begin{cases} (1+t)^{-1/(p-1)+1/2q} & \text{if} \quad 1 < p < p_F(1), \\ (1+t)^{-1/2+1/2q} \log(1+t)^{-1/2} & \text{if} \quad p = p_F(1) \end{cases} \tag{10}$$

with $1 \leq q \leq \infty$ for small initial data. This expectation comes from the asymptotic behavior of solutions:

$$\begin{cases} \partial_t v - \Delta v + v^p = 0, & (t, x) \in \mathbb{R} \times \mathbb{R}, \\ v(0, x) \geq 0, & x \in \mathbb{R}. \end{cases} \tag{11}$$

Indeed, in the subcritical case $1 < p < p_F(1)$, when $q = \infty$, the decay rate (10) for v is obtained by the ODE argument. Namely, $f(t) = ((p-1)t + 1)^{-1/(p-1)}$ satisfies

$$\partial_t f + f^p = 0.$$

Therefore f is a super solution of (11) as long as $\|v(0)\|_{L^\infty} \leq 1$ and the comparison argument implies (10) for v and $q = \infty$. For solutions to (11), the asymptotic equivalence

$$v(t, x) \sim \begin{cases} t^{-1/(p-1)} f(\frac{x}{\sqrt{t}}) & \text{if} \quad 1 < p < p_F(n), \\ ((1+t)\log(1+t))^{-n/2} f(\frac{x}{\sqrt{t}}) & \text{if} \quad p = p_F(n), \end{cases} \tag{12}$$

was shown [2, 3, 7] under some initial conditions, with radial self-similar profile f. Since the linear solution to (9) is asymptotically equivalent to that of (4), the decay estimate (10) is a natural expectation coming from (12).

Equation (10) has been obtained under some technical assumptions. Hayashi et al. [9, 10] showed that in the case where $p_1(F) - \varepsilon^2 < p \le p_1(F)$, the solution w to (9) satisfies (10) under the following initial condition: if $p < p_1(F)$, then

$$(u_0, (1+\partial_x)^{-1}u_1) = (\varepsilon f_0, \varepsilon f_1), \in (L^\infty \cap \langle x\rangle^{-a}L^1)^2$$

satisfies the initial momentum condition,

$$\int_{\mathbb{R}} (f_0(x) + f_1(x))dx > 0, \tag{13}$$

with $a > 0$ and sufficiently small $\varepsilon > 0$ and if $p = p_1(F)$,

$$(u_0, u_1) = (\varepsilon f_0, \varepsilon f_1), \in (H^{\delta,0} \cap H^{0,\delta}) \times (H^{\delta-1} \cap H^{\delta-1,\delta})$$

satisfies (13) and

$$\int_{\mathbb{R}} f_0(x)dx > 0,$$

where $\delta > 1/2$ and $H^{s,b}$ is defined by $(1-\Delta)^{-s/2}(1+|x|^2)^{-b/2}L^2$. We note that in [9, 10], not only the decay estimate (10) but also the asymptotic behavior (9) was shown. In [10], the two and three dimensional cases are treated as well. Moreover, in the general spatial dimensional case, Nishihara and Zhao [23] and Ikehata et al. [16] showed the decay estimate (10) by the modified energy method under the initial condition

$$\|(1+|x|)^m(u_0, \nabla u_0, u_1, |u_0|^{\frac{p+1}{2}})\|_{L^2} < \infty$$

with

$$m > \frac{2}{p-1} - \frac{n}{2}.$$

We also refer [26] and references therein for the related topics. In summary, the decay estimate (10) has been obtained with initial data in weighted Lebesgue or Sobolev spaces.

The aim of this manuscript is to obtain (10) by using the initial sign condition (2) but without any condition of weight. The main assertion of this manuscript is the following:

Theorem 1.2 *Let* $1 < p < 3$, $u_0 \in C^2 \cap W^{2,1} \cap W^{2,\infty}$ *and* $u_1 \in C^1 \cap L^1 \cap L^\infty$ *satisfy* (2). *If* $p > 5/3$, *then the corresponding solution* u *to* (1) *satisfies*

$$\sup_{t \geq 0}(1+t)^{1/(p-1)}\|u(t)\|_{L^\infty} < \infty.$$

If $p \leq 5/3$, *then there exist increasing sequences* $\{A_k\}_{k\geq 1}$ *and* $\{\alpha_k\}_{k\geq 1}$ *such that*

$$\lim_{k\to\infty} \alpha_k = 3/2,$$

$$\lim_{k\to\infty} A_k = \infty,$$

$$\|u(t)\|_{L^\infty} \leq A_k(1+t)^{-\alpha_k}.$$

Remark 1.1 *The threshold* $5/3$ *is technical. When* $p = 5/3$,

$$\frac{1}{p-1} = \frac{3}{2}.$$

Therefore, the L^∞ *decay rate obtained with Theorem 1.2 is at most* $t^{-3/2}$.

Theorem 1.2 is proved by the application of the comparison argument for the heat equation:

$$\begin{cases} \partial_t v - \Delta v + v^p = \partial_t^2 u, \\ v(0,x) = -u_0. \end{cases}$$

We note that if $p > 5/3$ and u behaves as the solution of (4) but v behaves as the solution of (11), then $\partial_t^2 u$ decays faster than u^p or v^p. Therefore, $\partial_t^2 u$ is expected to be regarded a negligible perturbation and the simple comparison argument of ODE implies Theorem 1.2. While when $p \leq 5/3$, under the same expectation, the decay rate of $\partial_t^2 u$ is slower than the nonlinearity. Therefore, when $p \leq 5/3$, the sharp decay rate of u cannot be obtained by this argument. For details, see the third section. In the next section, we collect some preliminary estimates and study a recurrent relation for the decay rates of u, $\partial_t u$, and $\partial_t^2 u$.

2 Preliminary

Let us remind some properties of linear solutions at first. Let I_j is the j-th modified Bessel function. Let $S(t)$ be a solution operator for linear damped wave equation defined by

$$S(t)f(x) = \frac{e^{-t/2}}{2}\int_{-t}^{t} I_0(\frac{\sqrt{t^2-y^2}}{2})f(x-y)dy.$$

For details, see [1, Chapter 6 Section 5 Part 7]. We rewrite $S(t)$ and its derivatives by

$$S(t)f(x) = \widetilde{\mathcal{K}}_0(t) * f(x), \tag{14}$$

$$\partial_t S(t)f(x) = e^{-t/2}\frac{f(x+t)+f(x-t)}{2} + \widetilde{\mathcal{K}}_1(t) * f(x), \tag{15}$$

$$\begin{aligned}\partial_t^2 S(t)f(x) &= e^{-t/2}\frac{f'(x+t)-f'(x-t)}{2} \\ &\quad + e^{-t/2}\left(\frac{t}{16}-\frac{1}{2}\right)(f(x+t)+f(x-t)) + \widetilde{\mathcal{K}}_2(t) * f(x),\end{aligned} \tag{16}$$

where

$$\mathcal{K}_0(t,y) = \frac{1}{2}e^{-t/2}I_0(\frac{\omega}{2}),$$

$$\mathcal{K}_1(t,y) = \frac{1}{4}e^{-t/2}\left(\frac{t}{\omega}I_1(\frac{\omega}{2}) - I_0(\frac{\omega}{2})\right),$$

$$\mathcal{K}_2(t,y) = e^{-t/2}\left(\frac{t^2}{16\omega^2}I_2(\frac{\omega}{2}) - \left(\frac{t}{4\omega}+\frac{y^2}{4\omega^3}\right)I_1(\frac{\omega}{2}) + \left(\frac{1}{8}+\frac{t^2}{16\omega^2}\right)I_0(\frac{\omega}{2})\right),$$

with $\omega = \sqrt{t^2-y^2}$ and

$$\widetilde{\mathcal{K}}_j(t,y) = \begin{cases}\mathcal{K}_j(t,y) & \text{if} \quad |y| \le t, \\ 0 & \text{if} \quad |y| > t\end{cases}$$

for $0 \le j \le 2$. For $0 \le j \le 2$, $\mathcal{K}_j$ is estimated as follows.

Lemma 2.1 ([21, Theorem 1.2]) *For $|y| \le t$ and $0 \le j \le 2$, we have*

$$|\mathcal{K}_j(t,y)| \le C\frac{e^{-y^2/8t}}{\langle t\rangle^{j+1/2}}.$$

In particular, for $1 \le q \le \infty$ and $0 \le j \le 2$, the following estimates hold:

$$\|\widetilde{\mathcal{K}}_j(t) * h\|_{L^q} \le C\langle t\rangle^{-j-1/2q'}\|h\|_{L^1}.$$

Then we estimate $\partial_t u$ and $\partial_t^2 u$. Here we put,

$$N(u)(t) = \int_0^t S(t-\tau)|u(\tau)|^p d\tau.$$

Lemma 2.2 *Let $p > 1$ and $\alpha > 0$. Let $u \in L^p([0,\infty) \times \mathbb{R}) \cap C^1((0,\infty); L^\infty)$ satisfy*

$$\|u(t)\|_{L^\infty} \le C(1+t)^{-\alpha}.$$

Then,

$$\begin{aligned}
\|\partial_t N(u)(t)\|_{L^\infty} &\le C(1+t)^{-3/2} + C(1+t)^{-\alpha p}\log(1+t),\\
\|\partial_t^2 N(u)(t)\|_{L^\infty} &\le C(1+t)^{-5/2}\\
&\quad + C(1+t)^{-\alpha(p-1)}\Big((1+t)^{-\alpha-1} + \|\partial_t u\|_{L^\infty([t/2,t]\times\mathbb{R})}\log(1+t)\Big).
\end{aligned}$$

Proof We compute

$$\partial_t N(u)(t) = \int_0^t \partial_t S(t-\tau)|u(\tau)|^p d\tau.$$

Lemma 2.1 implies

$$\begin{aligned}
&\|\partial_t N(t)\|_{L^\infty}\\
&\le C(1+t)^{-3/2}\int_0^{t/2} \|u(\tau)\|_{L^p}^p d\tau + C\int_{t/2}^t (1+t-\tau)^{-1}\|u(\tau)\|_{L^\infty}^p d\tau\\
&\le C\|u\|_{L^p([0,\infty)\times\mathbb{R}}^p (1+t)^{-3/2} + C(1+t)^{-\alpha p}\log(1+t).
\end{aligned}$$

Moreover, by changing the variable as $\sigma = t-\tau$ for $t/2 \le \tau \le t$, we have

$$\begin{aligned}
\partial_t N(t) &= \int_0^{t/2} \partial_t S(t-\tau)|u(\tau)|^p d\tau + \int_0^{t/2} \partial_t S(\sigma)|u(t-\sigma)|^p d\sigma,\\
\partial_t^2 N(t) &= \partial_t S(t/2)|u(t/2)|^p + \int_0^{t/2} \partial_t^2 S(t-\tau)|u(\tau)|^p d\tau\\
&\quad + p\int_0^{t/2} \partial_t S(\sigma)|u(t-\sigma)|^{p-2}u(t-\sigma)\partial_t u(t-\sigma)d\sigma.
\end{aligned}$$

Then, Lemma 2.1 implies

$$\begin{aligned}
\|\partial_t S(t)|u(t)|^p\|_{L^\infty} &\le C(1+t)^{-\alpha p-1},\\
\left|\int_0^{t/2} \partial_t^2 S(t-\tau)|u(\tau)|^p d\tau\right| &\le C(1+t)^{-5/2}\|u\|_{L^p([0,\infty)\times\mathbb{R})}^p,
\end{aligned}$$

$$\left| \int_0^{t/2} \partial_t S(\sigma) |u(t-\sigma)|^{p-2} u(t-\sigma) \partial_t u(t-\sigma) d\sigma \right| \le C(t+1)^{-(p-1)\alpha} \times \|\partial_t u\|_{L^\infty([t/2,t]\times\mathbb{R})} \log(1+t).$$

□

Now we consider the recurrence relation below. Since we estimate the decay rate of $\partial_t^2 u$ based on that of u and estimate the decay rate of u based on that of $\partial_t^2 u$ again, the recurrence relation is used to estimate the best decay rate of u.

Lemma 2.3 *Let* $1 < p < 3$ *and* $\alpha_1 = 1/2$*. For* $k \ge 1$*, we put*

$$\begin{aligned} \delta_k &= \min\{3/4 - \alpha_k/2, (p-1)/5\}, \\ \beta_k &= \min\{3/2, p\alpha_k - \delta_k\}, \\ \gamma_k &= \min\{5/2, p\alpha_k + 1, (p-1)\alpha_k + \beta_k - \delta_k\}, \\ \alpha_{k+1} &= \min\{1/(p-1), \gamma_k/p\}. \end{aligned}$$

If $p > 5/3$*, then there exists* k_0 *such that* $\alpha_{k_0} = 1/(p-1)$ *and if* $p \le 5/3$*, then* $\{\alpha_k\}_k$ *is an increasing sequence and whose limit is* $3/2$*.*

Proof We, at first, note that the identity

$$\gamma_k = (p-1)\alpha_k + \beta_k - \delta_k$$

holds because $\sup_k \alpha_k \le 1/(p-1)$ implies that

$$(p-1)\alpha_k + \beta_k \le 1 + \min\{3/2, p\alpha_k\}.$$

If $\alpha_k < 3/2$, then

$$\alpha_{k+1} \le \frac{\gamma_k}{p} \le \frac{(p-1)\alpha_k + 3/2}{p} < \frac{3}{2}.$$

Moreover, we have

$$\begin{aligned} \frac{(p-1)\alpha_k + 3/2}{p} - \alpha_k &= \frac{1}{p}\left(\frac{3}{2} - \alpha_k\right), \\ \frac{(2p-1)\alpha_k}{p} - \alpha_k &= \frac{p-1}{p}\alpha_k. \end{aligned}$$

These identities imply that $\alpha_{k+1} - \alpha_k$ is computed by

$$\alpha_{k+1} - \alpha_k \ge \min\left\{\frac{1}{p-1} - \alpha_k, \frac{1}{p}\left(\frac{3}{2} - \alpha_k - \delta_k\right), \frac{1}{p}\Big((p-1)\alpha_k - 2\delta_k\Big)\right\}.$$

If $p > 5/3$ and $\alpha_k \geq \alpha_1$, then we have

$$\alpha_{k+1} - \alpha_k \geq \min\left\{\frac{1}{p-1} - \alpha_k, \frac{1}{2p}\left(\frac{3}{2} - \alpha_k\right), \frac{p-1}{p}(\alpha_k - \frac{2}{5})\right\}$$
$$\geq \min\left\{\frac{1}{p-1} - \alpha_k, \frac{1}{2p}\left(\frac{3}{2} - \frac{1}{p-1}\right), \frac{p-1}{10p}\right\}.$$

Therefore, there exists k_0 such that $\alpha_{k_0} = 1/(p-1)$.

If $p \leq 5/3$ and $\alpha_k \geq \alpha_1$, then we estimate

$$\alpha_{k+1} - \alpha_k \geq \min\left\{\frac{1}{2p}\left(\frac{3}{2} - \alpha_k\right), \frac{p-1}{10p}\right\}.$$

Therefore, $\{\alpha_k\}_k$ is an increasing sequence whose limit is $3/2$. □

3 Proof

Now we consider the following heat equation:

$$\begin{cases} \partial_t v - \Delta v + |v|^{p-1}v = \partial_t^2 u, & (t,x) \in \mathbb{R} \times \mathbb{R}, \\ v(0,x) = -u_0, & x \in \mathbb{R}. \end{cases} \tag{17}$$

The existence of classical C^2 solution to (1) is guaranteed by the assumptions of u_0, u_1. It is easy to show that $u + v = 0$. Indeed, we can use the fact that $u + v$ satisfies

$$\begin{cases} \partial_t(u+v) - \Delta(u+v) = (-u)^p - |v|^{p-1}v, & (t,x) \in \mathbb{R} \times \mathbb{R}, \\ (u+v)(0,x) = 0, & x \in \mathbb{R}. \end{cases}$$

A standard contraction argument implies that $v = -u > 0$. Since v satisfies the comparison principle, if $f \in C([0,\infty); \mathbb{R}) \cap C^1((0,\infty); \mathbb{R})$ satisfies

$$\begin{cases} f' + f^p - \|\partial_t^2 u\|_{L^\infty} \geq 0, & t > 0, \\ f(0) \geq \|u_0\|_{L^\infty}, \end{cases} \tag{18}$$

then $\|v(t)\|_{L^\infty} \leq f(t)$ for $t \geq 0$. According to Proposition 1.1, Lemma 2.2 with $\alpha_1 = 1/2$ implies that with a proper $\delta_1 > 0$, the estimates

$$\|\partial_t u(t)\|_{L^\infty} \leq C(1+t)^{-\beta_1},$$
$$\|\partial_t^2 u(t)\|_{L^\infty} \leq C(1+t)^{-\gamma_1}$$

hold. If we put $f_{\alpha_2} = A_2(t+1)^{-\alpha_2}$, then with α_2 determined by Lemma 2.3 with $\alpha_1, \beta_1, \gamma_1, \delta_1$, the estimate

$$\begin{aligned} f'_{\alpha_2} + f^p_{\alpha_2} - \|\partial_t^2 u\|_{L^\infty} &= -A_2\alpha_2(t+1)^{-\alpha_2-1} + A_2^p(t+1)^{-p\alpha_2} - C(1+t)^{-\gamma_1} \\ &\geq \frac{A_2^p}{2}(1+t)^{-p\alpha_2} > 0. \end{aligned}$$

holds with sufficiently large A_2. Then f_{α_2} satisfies (18) and

$$\|u(t)\|_{L^\infty} \leq A_2(1+t)^{-\alpha_2}.$$

Repeating this argument with Lemma 2.2 and the comparison argument, we define $f_{\alpha_k} = A_k(t+1)^{-\alpha_k}$ with $k \geq 2$ and A_k such that f_{α_k} is a super solution of (18). Then we have

$$\|u(t)\|_{L^\infty} \leq A_k(1+t)^{-\alpha_k}$$

for any $k \geq 2$. Therefore, if $p > 5/3$, the estimate

$$\|u(t)\|_{L^\infty} \leq C(1+t)^{-1/(p-1)}$$

holds with sufficiently large C. If $p \leq 5/3$, then the estimate

$$\|u(t)\|_{L^\infty} \leq A_k(1+t)^{-\alpha_k}$$

holds with increasing sequences $\{A_k\}_k$ and $\{\alpha_k\}_k$ whose limit is $3/2$.

Acknowledgments The authors are grateful to the referees for their careful reading and helpful comments.

The first author was supported in part by JSPS Grant-in-Aid for Early-Career Scientists No. 20K14337 and 24K16957. The second author was partially supported by Gruppo Nazionale per l'Analisi Matematica, project "Modelli nonlineari in presenza di interazioni puntuali" CUP—E53C22001930001, by the project PRIN 2020XB3EFL with the Italian Ministry of Universities and Research, by Institute of Mathematics and Informatics, Bulgarian Academy of Sciences, by Top Global University Project, Waseda University and the Project PRA 2022 85 of University of Pisa.

References

1. Courant, R., Hilbert, D.: Methods of Mathematical Physics: Partial Differential Equations, volume 2 of Methods of Mathematical Physics. Wiley, New York (1989)
2. Escobedo, M., Kavian, O.: Asymptotic behaviour of positive solutions of a nonlinear heat equation. Houston J. Math. **14**(1), 39–50 (1988)

3. Escobedo, M., Kabian, O., Matano, H.: Large time behavior of solutions of a dissipative semilinear heat equation. Commun. Partial Differ. Equations **20**(7–8), 1427–1452 (1995)
4. Fujita, H.: On the blowing up of solutions of the Cauchy problem for $u_t = \Delta u + u^{1+\alpha}$. J. Fac. Sci. Univ. Tokyo Sect. I **13**, 109–124 (1966)
5. Fujiwara, K., Georgiev, V.: On extended lifespan for 1d damped wave equation. Preprint.
6. Fujiwara, K., Ikeda, M., Wakasugi, Y.: Estimates of lifespan and blow-up rates for the wave equation with a time-dependent damping and a power-type nonlinearity. Funkcial. Ekvac. **62**(2), 157–189 (2019)
7. Galaktionov, V.A., Kurdyumov, S.P., Samarskll, A.A.: On asymptotic "eigenfunctions" of the cauchy problem for a nonlinear parabolic equation. In: C6opHHK Math. USSR Sbornik TOM, vol. 126 (1985)
8. Hayakawa, K.: On nonexistence of global solutions of some semilinear parabolic differential equations. Proc. Japan Acad. **49**, 503–505 (1973)
9. Hayashi, N., Kaikina, E.I., Naumkin, P.I.: Damped wave equation in the subcritical case. J. Differ. Equations **207**, 161–194 (2004)
10. Hayashi, N., Kaikina, E., Naumkin, P.: Damped wave equation with a critical nonlinearity. Trans. Am. Math. Soc. **358**, 1165–1185 (2005)
11. Ikeda, M., Ogawa, T.: Lifespan of solutions to the damped wave equation with a critical nonlinearity. J. Differ. Equations **261**(3), 1880–1903 (2016)
12. Ikeda, M., Sobajima, M.: Sharp upper bound for lifespan of solutions to some critical semilinear parabolic, dispersive and hyperbolic equations via a test function method. Nonlinear Anal. **182**, 57–74 (2019)
13. Ikeda, M., Wakasugi, Y.: A note on the lifespan of solutions to the semilinear damped wave equation. Proc. Amer. Math. Soc. **143**(1), 163–171 (2015)
14. Ikeda, M., Inui, T., Okamoto, M., Wakasugi, Y.: $L^p - L^q$ estimates for the damped wave equation and the critical exponent for the nonlinear problem with slowly decaying data. Commun. Pure Appl. Anal. **18**(4), 1967–2008 (2019)
15. Ikehata, R., Tanizawa, K.: Global existence of solutions for semilinear damped wave equations in RN with noncompactly supported initial data. Nonlinear Anal. Theory Methods Appl. **61**(7), 1189–1208 (2005)
16. Ikehata, R., Nishihara, K., Zhao, H.: Global asymptotics of solutions to the cauchy problem for the damped wave equation with absorption. J. Differ. Equations **226**, 1–29 (2006)
17. Karch, G.: Selfsimilar profiles in large time asymptotics of solutions to damped wave equations. Studia Math. **143**, 175–197 (2000)
18. Kawakami, T., Takeda, H.: Higher order asymptotic expansions to the solutions for a nonlinear damped wave equation. Nonlinear Differ. Equations Appl. NoDEA **23**, 54 (2016)
19. Kobayashi, K., Sirao, T., Tanaka, H.: On the growing up problem for semilinear heat equations. J. Math. Soc. Japan **29**(3), 407–424 (1977)
20. Li, T.T., Zhou, Y.: Breakdown of solutions to $\Box u + u_t = |u|^{1+\alpha}$. Discrete Contin. Dyn. Syst. **1**(4), 503–520 (1995)
21. Marcati, P., Nishihara, K.: The Lp 2 Lq estimates of solutions to one-dimensional damped wave equations and their application to the compressible flow through porous media. J. Differ. Equations **191**, 445–469 (2003)
22. Nishihara, K.: Lp-Lq estimates of solutions to the damped wave equation in 3-dimensional space and their application. Math. Z. **244**(3), 631–649 (2003)
23. Nishihara, K., Zhao, H.: Decay properties of solutions to the cauchy problem for the damped wave equation with absorption. J. Math. Anal. Appl. **313**, 598–610 (2006)
24. Strauss, W.A.: Nonlinear Wave Equations. Number 73 in CBMS Regional Conference Series. Conference Board of the Mathematical Sciences (1990)
25. Todorova, G., Yordanov, B.: Critical exponent for a nonlinear wave equation with damping. J. Differ. Equations **174**(2), 464–489 (2001)
26. Wakasugi, Y.: Decay property of solutions to the wave equation with space-dependent damping, absorbing nonlinearity, and polynomially decaying data. Math. Methods Appl. Sci. **46**(6), 7067–7107 (2023)

Asymptotic Behavior for the Massive Maxwell-Klein-Gordon System Under the Lorenz Gauge Condition in Dimension (1+4)

Kiyeon Lee

Abstract In this paper, we consider the Cauchy problem for the *massive* Maxwell-Klein-Gordon system (MKG) under the Lorenz gauge condition in (4+1) dimensional Minkowski space. Due to the parallelism to the Maxwell-Dirac system, we provide only estimates that differ from those in the Maxwell-Dirac system with Maxwell-Dirac system with those in the previous work (arXiv:2312.13621).

Keywords Maxwell-Klein-Gordon system · Global existence · Space-time resonance · Scattering theory · Lorenz gauge condition

2020 Mathematics Subject Classification 35Q41, 35Q55, 35Q40

1 Introduction

1.1 Maxwell-Klein-Gordon System

The (1+4)-dimensional Maxwell-Klein-Gordon system is given by

$$\begin{cases} \mathbf{D}^{\mu}\mathbf{D}_{\mu}\phi = -m^2\phi, \\ \partial^{\nu}F_{\mu\nu} = J_{\mu} \end{cases} \tag{MKG}$$

where the scalar field is $\phi : \mathbb{R}^{1+4} \to \mathbb{C}$, the gauge fields are $A_\mu : \mathbb{R}^{1+4} \to \mathbb{R}$ and the covariant derivative is $\mathbf{D}_\mu = \partial_\mu + iA_\mu$ for $\mu = 0, \cdots, 4$. The curvature is defined by $F_{\mu\nu} = \partial_\mu A_\nu - \partial_\nu A_\mu$. The source term is defined by $J_\mu = \mathrm{Im}\left(\phi\overline{\mathbf{D}_\mu\phi}\right)$. Greek indices indicate the space-time components $\mu, \nu = 0,\ 1,\ 2,\ 3,\ 4$ and roman indices

K. Lee (✉)
Department of Mathematical Sciences, Korea Advanced Institute of Science and Technology, Daejeon, Republic of Korea
e-mail: kiyeonlee@kaist.ac.kr

V. S. Georgiev, T. Ozawa (eds.), *Various Methods for the Analysis of PDEs*,
ICIAM2023 Springer Series 5, https://doi.org/10.1007/978-981-96-9164-7_2

mean the spatial components $j = 1, 2, 3, 4$ in the sequel. The Einstein summation convention is in effect with Greek indices summed over $\mu = 0, \cdots, 4$ and Latin indices summed over the spatial variables $j = 1, \cdots, 4$. Thus $\partial^\mu = \eta^{\mu\nu}\partial_\nu$, and $\partial_0 = \partial_t$ with Minkowski metric $\eta = \mathrm{diag}(-1, 1, \cdots, 1)$. We call *massive* and *massless* (MKG) if the mass parameter is $m > 0$ and $m = 0$, respectively. This paper only considers the *massive* case. Thus, we normalize the mass of the scalar field to be 1.

The one of basic features of (MKG) is the gauge invariance. Indeed, (MKG) is invariant under the gauge transformation $(\phi, A) \longmapsto (e^{i\chi}\phi, A - d\chi)$, for a real-valued function χ on $\mathbb{R} \times \mathbb{R}^4$. For the sake of concreteness of our discussion, let us choose Lorenz gauge, which is defined by the condition

$$\partial^\mu A_\mu = 0. \tag{1.1}$$

Under Lorenz gauge, (MKG) becomes

$$\begin{cases} (\Box + 1)\phi = i\partial^\mu \left(A_\mu\phi\right) + iA^\mu\partial_\mu\phi - A^\mu A_\mu\phi, \\ \Box A_\mu = -\mathrm{Im}\left(\phi\overline{\partial_\mu\phi}\right) - A_\mu|\phi|^2, \end{cases} \quad \text{in } \mathbb{R}^{1+4} \tag{1.2}$$

with initial data

$$\phi(0) = \phi_0, \quad \partial_t\phi(0) = \dot{\phi}_0 \quad A_\mu(0) = a_\mu, \quad \partial_t A_\mu(0) = \dot{a}_\mu. \tag{1.3}$$

We denoted $\Box = -\partial_t^2 + \Delta$. For other common choices, there are Coulomb gauge condition $\partial^j A_j = 0$ and temporal gauge condition $A_0 = 0$.

1.2 Previous Works

There is a large amount of literature dealing with the problem of local and global well-posedness and asymptotic behavior of small solutions of (MKG).

In the early work [6, 7], the authors investigated the local and global existence of the (1+3) dimensional Yang-Mills-Higgs system in $H^2 \times H^1 \times H^2 \times H^1$. Klainerman and Machedon [20] showed the global existence and regularity result in the energy space by using the null forms for the *massless* case. See also [18, 19]. For the global well-posedness of the (1+4) dimensional *massless* wave equation (MKG), we also refer to [23, 29]. Independent studies on the scattering phenomena of *massless* (MKG) in (1+4) Minkowski space under the Coulomb gauge condition were conducted in [22, 25, 26].

Regarding the global existence in the (1+3)-dimensional Minkowski space, we can refer to [30]. The papers [2, 31, 32] analyzed the asymptotic behavior of solutions to the *massless* (MKG) under both Coulomb and Lorenz gauge conditions in (1+3) dimensions.

While extensive attention has been given to *massless* wave equations by various authors, fewer studies have focused on the scattering results for *massive* counterparts. Notably, the recent research in [9] established the global regularity for modified scattering results of the *massive* (MKG) under the Coulomb gauge condition in (1+4) dimensions. In [15], the authors exploited the Fourier type method with the vector-field energy estimates in order to obtain global existence and modified scattering results for (1+3) dimensional wave-Klein-Gordon equations. This approach may also apply to (MKG). Additionally, see [8, 21] and [4, 5] for investigations into the asymptotic behavior of the (1+3)-dimensional *massive* (MKG) and related wave-Klein-Gordon type systems. Recently, the modified scattering results for the (1+3)-dimensional *massive* Maxwell-Dirac system have been established in [3, 14].

1.3 Main Theorem

While some modified scattering results are known, the *linear* scattering results of both *massive* and *massless* (MKG) are unknown in any dimension whatever we impose any gauge conditions. In this paper, we consider the global existence of the solution to (1.2) and scattering results for *massive* (1.2) ($m > 0$) on Minkowski space $\mathbb{R}^{1+4}$. For the sake of simplicity, we normalize $m = 1$. Our main result is the following:

Theorem 1.1 *Let n is sufficiently large number. Then we have the following.*

(1) Let initial data (1.3) *satisfy that, for some $\varepsilon_0 > 0$,*

$$\|(\phi_0, \dot{\phi}_0)\|_{H^n \times H^{n-1}} + \| \langle x \rangle^2 \phi_0 \|_{L^2} + \|(a_\mu, \dot{a}_\mu)\|_{H^n \times H^{n-1}} + \sum_{k=1,2} \|(x^k a_\mu, x^k \dot{a}_\mu)\|_{\dot{H}^k \times \dot{H}^{k-1}} < \varepsilon_0. \tag{1.4}$$

Then, there exists a global solution $(\phi(t), A_\mu(t))$ to (MKG) *under the Lorenz gauge condition* (1.1).

(2) Under the assumption (1.4), *the solution decays as follows:*

$$\|\phi(t)\|_{L^\infty} \lesssim \langle t \rangle^{-2+\delta} \varepsilon_0, \quad \textit{and} \quad \|A_\mu(t)\|_{L^\infty} \lesssim \langle t \rangle^{-\frac{3}{2}} \varepsilon_0, \tag{1.5}$$

for some small $\delta := \delta(n) > 0$.

(3) *Moreover, the solution* $(\phi(t), A_\mu(t))$ *to* (MKG) *scatters linearly in* $H^1 \times L^2 \times \dot{H}^1 \times L^2$ *as* $t \to \infty$. *Indeed, there exist* $(\phi^\infty(t), A_\mu^\infty(t))$ *such that*

$$\left\|(\phi(t), \partial_t\phi(t)) - (\phi^\infty(t), \partial_t\phi^\infty(t))\right\|_{H^1\times L^2} + \left\|(A_\mu(t), \partial_t A_\mu(t)) - (A_\mu^\infty(t), \partial_t A_\mu^\infty(t))\right\|_{\dot{H}^1\times L^2} \xrightarrow{t\to\infty} 0,$$

and $(\phi^\infty(t), A_\mu^\infty(t))$ *is a solution to the linear system* (MKG)*:*

$$\begin{cases} \Box\phi^\infty = \phi^\infty, \\ \Box A_\mu^\infty = 0. \end{cases}$$

Remark 1.2 In (1.5), the parameter δ is depending on regularity n. Suitably large number n implies the almost optimal rate of Klein-Gordon operator t^{-2+}. Note that Theorem 1.1 shows the optimal time decay of the wave operator.

1.4 Main Idea

From the parallel properties between (MKG) and Maxwell-Dirac system, we can apply the approaches established in the author's previous result [24]. In [24] the proof of global solution and scattering results relies on a bootstrap argument under suitable a priori assumptions with weighted energy estimates. To achieve this, the space-time resonance framework developed by Germain et al. [10–12] and [13] is exploited. By applying Duhamel's formula, the nonlinear terms of (1.2) lead to several oscillatory integrals, whose large-time behaviors are determined by the stationary points of the phase function. The analysis begins with addressing the space, time, and space-time resonance of *phases* originating from the Klein-Gordon spinor and wave operator within the oscillatory integrals, arising from Klein-Gordon and wave propagators. This type of resonances has been established in [16, 24, 27], we revisit the resonance sets. Specifically, in the nonlinearities of (MKG) the cubic terms can be estimated in an easier way, we only focus on the quadratic nonlinearities. Then, we obtain the system of the following nonlinearities $\mathbf{f}_\Theta$ and $\mathbf{F}_{\mu,K}$ for the Klein-Gordon and Maxwell parts, respectively:

$$\begin{aligned}
\mathbf{f}_\Theta(t,\xi) &:= \int_0^t\int_{\mathbb{R}^4} e^{isp_\Theta(\xi,\eta)}\frac{\xi^j + (\xi-\eta)^j}{\langle\xi\rangle}\widehat{F_{j,\theta_2}}(s,\eta)\widehat{f_{\theta_1}}(s,\xi-\eta)\,d\eta ds,\\
\mathbf{F}_{\mu,K}(t,\xi) &:= \int_0^t\int_{\mathbb{R}^4} e^{isq_K(\xi,\eta)}\frac{\xi_j + 2\eta_j}{|\xi|}\widehat{f_{\kappa_1}}(s,\xi+\eta)\overline{\widehat{f_{\kappa_2}}(s,\eta)}\,d\eta ds,\\
p_\Theta(\xi,\eta) &:= -\theta_0\langle\xi\rangle + \theta_1\langle\xi-\eta\rangle + \theta_2|\eta|,\\
q_K(\xi,\eta) &:= -\kappa_0|\xi| + \kappa_1\langle\xi+\eta\rangle - \kappa_2\langle\eta\rangle,
\end{aligned}$$

for 3-tuples $\Theta = (\theta_0, \theta_1, \theta_2)$ and $K = (\kappa_0, \kappa_1, \kappa_2)$ where $\theta_j, \kappa_j \in \{+, -\}$ $(j = 0, 1, 2)$ (see Sect. 2.1 for the rigorous derivation). The functions p_Θ and q_K are commonly referred to as *phases*. These phases exhibit various sign relations, requiring consideration of different resonance cases.

In this paper, the similar parts will be omitted and we will consider the different parts with those in [24]. Given the nonlinearities, since the cases that time derivative falls on wave functions should be handled under the a priori assumption, we examine the estimates for the cases. We also treat the energy estimates of cubic nonlinearities.

2 Preliminaries

Notations

(1) (Sobolev spaces) Let $s \in \mathbb{R}$. We define homogeneous Sobolev spaces $\|u\|_{\dot{H}^s} := \||D|^s u\|_{L^2}$. For inhomogeneous spaces, we define $\|u\|_{H^s} := \|\langle D\rangle^s u\|_{L^2}$. Moreover, $\|u\|_{W^{s,p}} := \|\langle D\rangle^s u\|_{L^p}$ for $p \neq 2$.

(2) (Littlewood-Paley operators) Let ρ be a Littlewood-Paley function such that $\rho \in C_0^\infty(B(0,2))$ with $\rho(\xi) = 1$ for $|\xi| \leq 1$ and define $\rho_N(\xi) := \rho\left(\frac{\xi}{N}\right) - \rho\left(\frac{2\xi}{N}\right)$ for $N \in 2^{\mathbb{Z}}$. Then we define the frequency projection P_N by $\mathcal{F}(P_N f)(\xi) = \rho_N(\xi)\widehat{f}(\xi)$, and also $\rho_{\leq N_0} := 1 - \sum_{N>N_0} \rho_N$. For $N \in 2^{\mathbb{Z}}$ we denote $\widetilde{\rho_N} = \rho_{N/2} + \rho_N + \rho_{2N}$. In particular, $\widetilde{P_N} P_N = P_N \widetilde{P_N} = P_N$ where $\widetilde{P_N} = \mathcal{F}^{-1}\widetilde{\rho_N}\mathcal{F}$. Especially, we denote $P_N f$ by f_N for any measurable function f.

2.1 *Set Up for Maxwell-Klein-Gordon System Under Lorenz Gauge Condition*

We decompose the scalar and gauge fields into half waves. To do this we define projection operators as follows:

$$\phi_\theta := \frac{1}{2}\left[1 + \theta \langle D\rangle^{-1}(-i\partial_t)\right]\phi,$$

$$A_{\mu,\kappa} := \frac{1}{2}\left[1 + \kappa|D|^{-1}(-i\partial_t)\right]A_\mu,$$

for $\theta, \kappa \in \{+,-\}$. Then, we decompose ϕ, A_μ into $\phi = \phi_+ + \phi_-$ and $A_\mu = A_{\mu,+} + A_{\mu,-}$, respectively. Therefore (1.2) becomes that

$$\begin{cases} (i\partial_t + \theta \langle D\rangle)\phi_\theta = -\theta\frac{1}{2}\langle D\rangle^{-1}\left(i\partial^\mu\left(A_\mu\phi\right) + iA^\mu\partial_\mu\phi - A^\mu A_\mu\phi,\right), \\ (i\partial_t + \kappa|D|)A_{\mu,\kappa} = -\kappa\frac{1}{2}|D|^{-1}\left(\mathrm{Im}\left(\phi\overline{\partial_\mu\phi}\right) - A_\mu|\phi|^2\right), \end{cases} \tag{2.1}$$

with initial data $(\phi_\theta(0), A_{\mu,\kappa}(0)) = (\phi_{0,\theta}, a_{\mu,\kappa})$, for $\theta, \kappa \in \{+,-\}$. We abbreviate the nonlinearity:

$$\mathfrak{N}_1^{kg} := i\partial^\mu(A_\mu\phi), \qquad \mathfrak{N}_2^{kg} := iA^\mu\partial_\mu\phi, \qquad \mathfrak{N}_3^{kg} := -A^\mu A_\mu\phi$$
$$\mathfrak{N}_1^{m} := \mathrm{Im}(\phi\overline{\partial_\mu\phi}), \qquad \mathfrak{N}_2^{m} := A_\mu|\phi|^2.$$

By Duhamel's principle, (2.1) can be converted into

$$\phi_{\theta_0}(t) = e^{\theta_0 it\langle D\rangle}\phi_{0,\theta_0} - \theta_0\frac{i}{2}\int_0^t e^{\theta_0 i(t-s)\langle D\rangle}\langle D\rangle^{-1}\left(\mathfrak{N}_1^{kg} + \mathfrak{N}_2^{kg} + \mathfrak{N}_3^{kg}\right)(s)\,ds, \tag{2.2}$$

$$A_{\mu,\kappa_0}(t) = e^{\kappa_0 it|D|}a_{\mu,\kappa_0} - \kappa_0\frac{i}{2}\int_0^t e^{\kappa_0 i(t-s)|D|}|D|^{-1}\left(\mathfrak{N}_1^{m} + \mathfrak{N}_2^{m}\right)(s)\,ds. \tag{2.3}$$

To observe the scattering state, we define propagator-adapted fields f_θ and $F_{\mu,\kappa}$ by

$$f_\theta(t) = e^{-\theta it\langle D\rangle}\phi_\theta(t) \quad \text{and} \quad F_{\mu,\kappa}(t) = e^{-\kappa it|D|}A_{\mu,\kappa}(t). \tag{2.4}$$

Then by acting propagators on both sides of (2.2) and (2.3), respectively, we have

$$f_{\theta_0}(t) = \phi_{0,\theta_0} - \theta_0\frac{i}{2}\int_0^t e^{-\theta_0 is\langle D\rangle}\langle D\rangle^{-1}\left(\mathfrak{N}_1^{kg} + \mathfrak{N}_2^{kg} + \mathfrak{N}_3^{kg}\right)(s)\,ds, \tag{2.5}$$

$$F_{\mu,\kappa_0}(t) = a_{\mu,\kappa_0} - \kappa_0\frac{i}{2}\int_0^t e^{-\kappa_0 is|D|}|D|^{-1}\left(\mathfrak{N}_1^{m} + \mathfrak{N}_2^{m}\right)(s)\,ds, \tag{2.6}$$

for $\theta_0, \kappa_0 \in \{+,-\}$. By taking Fourier transform, one gets the frequency representation as follows:

$$\widehat{f_{\theta_0}}(t,\xi) = \widehat{\phi_{0,\theta_0}} - \theta_0\frac{i}{2}\left[\sum_{\theta_1,\theta_2\in\{\pm\}}\left(\mathbf{f}_\Theta^{time}(t,\xi) + \mathbf{f}_\Theta^{quad}(t,\xi)\right) + \mathbf{f}_{\theta_0}^{cub}(t,\xi)\right],$$

$$\widehat{F_{\mu,\kappa_0}}(t,\xi) = \widehat{a_{\mu,\kappa_0}}(\xi) - \kappa_0\frac{i}{2}\left[\sum_{\theta_1,\theta_2\in\{\pm\}}\left(\mathbf{F}_{0,K}^{time}(t,\xi) + \mathbf{F}_{j,K}^{quad}(t,\xi)\right) + \mathbf{F}_{\mu,\kappa_0}^{cub}(t,\xi)\right],$$

where

$$
\begin{aligned}
\mathbf{f}_{\Theta}^{time}(t,\xi) &:= c_1 \int_0^t \int_{\mathbb{R}^4} e^{isp_\Theta(\xi,\eta)} \langle \xi \rangle^{-1} \left(\partial_s \widehat{F_{0,\theta_2}}(s,\eta) \widehat{f_{\theta_1}}(s,\xi-\eta) \right. \\
&\qquad \left. + 2\widehat{F_{0,\theta_2}}(s,\eta) \partial_s \widehat{f_{\theta_1}}(s,\xi-\eta) \right) d\eta ds, \\
\mathbf{f}_{\Theta}^{quad}(t,\xi) &:= c_2 \int_0^t \int_{\mathbb{R}^4} e^{isp_\Theta(\xi,\eta)} \frac{\xi^j + (\xi-\eta)^j}{\langle \xi \rangle} \widehat{F_{j,\theta_2}}(s,\eta) \widehat{f_{\theta_1}}(s,\xi-\eta)\, d\eta ds, \\
\mathbf{f}_{\theta_0}^{cub}(t,\xi) &:= c_3 \int_0^t \int_{\mathbb{R}^4} e^{-\theta_0 is\langle \xi \rangle} \langle \xi \rangle^{-1} \widehat{A^\mu A_\mu}(s,\eta) \widehat{\phi}(s,\xi-\eta)\, d\eta ds,
\end{aligned}
\tag{2.7}
$$

$$
\begin{aligned}
\mathbf{F}_{0,K}^{time}(t,\xi) &:= d_1 \mathrm{Im} \int_0^t \int_{\mathbb{R}^4} e^{isq_K(\xi,\eta)} |\xi|^{-1} \widehat{f_{\kappa_1}}(s,\xi+\eta) \overline{\partial_s \widehat{f_{\kappa_2}}(s,\eta)}\, d\eta ds, \\
\mathbf{F}_{j,K}^{quad}(t,\xi) &:= d_2 \int_0^t \int_{\mathbb{R}^4} e^{isq_K(\xi,\eta)} \frac{\xi_j + 2\eta_j}{|\xi|} \widehat{f_{\kappa_1}}(s,\xi+\eta) \overline{\widehat{f_{\kappa_2}}(s,\eta)}\, d\eta ds, \\
\mathbf{F}_{\mu,\kappa_0}^{cub}(t,\xi) &:= d_3 \int_0^t \int_{\mathbb{R}^4} e^{-\kappa_0 is|\xi|} \langle \xi \rangle^{-1} \widehat{A_\mu}(s,\xi-\eta) \widehat{|\phi|^2}(s,\eta)\, d\eta ds,
\end{aligned}
\tag{2.8}
$$

and

$$
p_\Theta(\xi,\eta) := -\theta_0 \langle \xi \rangle + \theta_1 \langle \xi - \eta \rangle + \theta_2 |\eta|, \tag{2.9}
$$

$$
q_K(\xi,\eta) := -\kappa_0 |\xi| + \kappa_1 \langle \xi + \eta \rangle - \kappa_2 \langle \eta \rangle, \tag{2.10}
$$

for 3-tuples $\Theta = (\theta_0, \theta_1, \theta_2)$ and $K = (\kappa_0, \kappa_1, \kappa_2)$. We call p_Θ, q_K the *phase*.

2.2 Space-Time Resonance for Maxwell-Klein-Gordon System

We proceed with our proof by a bootstrap argument based on the space-time resonance argument [10–12], and [13], which is our frame work. In this section, we investigate time, space, and space-time resonances for *massive* Maxwell-Klein-Gordon system, respectively. These resonance sets have been observed in [16, 17, 24, 27]. We also refer to [1, 28] for the analogous observation of the Klein-Gordon operator. The exact same calculation has been shown in [24]. However, since we will invoke the calculations in [24] in which the space-time resonance sets are exploited, we reexamine the space-time resonance sets for the (MKG) in this section. Thanks to the fact that MKG and MD share the space-time resonance sets, we may utilize the estimates in [24]. According to the definition in [10], we define time, space, and space-time resonant sets as follows:

$$\mathcal{T}_{p_\Theta} := \{(\xi,\eta) : p_\Theta(\xi,\eta) = 0\}, \qquad \mathcal{T}_{q_K} := \{(\xi,\eta) : q_K(\xi,\eta) = 0\},$$
$$\mathcal{S}_{p_\Theta} := \left\{(\xi,\eta) : \nabla_\eta p_\Theta(\xi,\eta) = 0\right\}, \qquad \mathcal{S}_{q_K} := \left\{(\xi,\eta) : \nabla_\eta q_K(\xi,\eta) = 0\right\},$$
$$\mathcal{R}_{p_\Theta} := \mathcal{T}_{p_\Theta} \cap \mathcal{S}_{p_\Theta}, \qquad \mathcal{R}_{q_K} := \mathcal{T}_{q_K} \cap \mathcal{S}_{q_K}.$$

Then we provide the results of calculations of them. We have

$$|p_\Theta(\xi,\eta)| \gtrsim \begin{cases} \frac{|\eta|}{\langle\xi\rangle(\langle\xi\rangle+\langle\xi-\eta\rangle+\langle\eta\rangle)} & \text{for } \theta_1 = \theta_2, \\ \langle\xi\rangle^{-1} & \text{otherwise,} \end{cases}$$

Therefore $\mathcal{T}_{p_\Theta} = \{\eta = 0\}$ when $\theta_0 = \theta_1$ and $\mathcal{T}_{p_\Theta} = \emptyset$, otherwise. Similarly, for the space resonance, we have

$$\left|\nabla_\eta p_\Theta(\xi,\eta)\right| \gtrsim \langle\xi-\eta\rangle^{-2}.$$

Therefore we deduce that $\mathcal{S}_{p_\Theta} = \emptyset$, for any sign relations, which implies $\mathcal{R}_{p_\Theta} = \emptyset$. By virtue of a similarity between $p_\Theta(\xi,\eta)$ and $q_K(\xi,\eta)$, we readily see that

$$|q_K(\xi,\eta)| \gtrsim \begin{cases} \frac{|\xi|}{\langle\eta\rangle(\langle\xi\rangle+\langle\xi-\eta\rangle+\langle\eta\rangle)} & \text{for } \kappa_1 = \kappa_2, \\ \langle\eta\rangle^{-1} & \text{otherwise,} \end{cases}$$

and the direct calculation implies that

$$|\nabla_\eta q_K(\xi,\eta)| \gtrsim \begin{cases} \frac{|\xi|}{\langle\xi+\eta\rangle^3} & \text{for } \kappa_1 = \kappa_2, \\ \left|\frac{\xi+\eta}{\langle\xi+\eta\rangle} + \frac{\eta}{\langle\eta\rangle}\right| & \text{otherwise.} \end{cases}$$

Note that space resonance of q_K is different from that of p_Θ. Therefore, we have $\mathcal{T}_{q_K} = \mathcal{S}_{q_K} = \mathcal{R}_{q_K} = \{\xi = 0\}$ when $\kappa_1 = \kappa_2$. On the other hand, when $\kappa_1 \neq \kappa_2$, $\mathcal{T}_{q_K} = \emptyset$, $\mathcal{S}_{q_K} = \{\xi = 2\eta\}$, and $\mathcal{R}_{q_K} = \emptyset$.

2.3 A Priori Assumption

As we discussed in introduction, we set a priori assumption in this section for the bootstrap argument. For $\varepsilon_1 > 0$ to be chosen later, we assume a priori smallness of solutions: for some $0 < \beta < \delta$, a given time $T > 0$, a priori assumptions of Klein-Gordon part and Maxwell part are suggested by

$$\|\phi\|_{\Sigma_T^{\mathbf{D}}} + \|A_\mu\|_{\Sigma_T^{\mathbf{M}}} \lesssim \varepsilon_1, \tag{2.11}$$

where

$$\|\phi\|_{\Sigma_T^{\mathbf{D}}} := \|\phi_+\|_{\Sigma_{T,+}^{\mathbf{D}}} + \|\phi_-\|_{\Sigma_{T,-}^{\mathbf{D}}}$$

$$\|\phi_\theta\|_{\Sigma_{T,\theta}^{\mathbf{D}}} := \sup_{t\in[0,T]} \left[\|\phi_\theta(t)\|_{H^n} + \|xf_\theta(t)\|_{L^2} + \langle t\rangle^{-\delta+\beta} \left\| x^2 f_\theta(t) \right\|_{L^2} \right],$$

and

$$\|A_\mu\|_{\Sigma_T^{\mathbf{M}}} := \|A_{\mu,+}\|_{\Sigma_{T,+}^{\mathbf{M}}} + \|A_{\mu,-}\|_{\Sigma_{T,-}^{\mathbf{M}}}$$

$$\|A_{\mu,\kappa}\|_{\Sigma_{T,\kappa}^{\mathbf{M}}} := \sup_{t\in[0,T]} \Bigg[\langle t\rangle^{-\frac{\delta}{2}} \|A_{\mu,\kappa}(t)\|_{H^n} + \langle t\rangle^{-\frac{\delta}{2}} \left\|xF_{\mu,\kappa}(t)\right\|_{\dot{H}^1} + \langle t\rangle^{-\frac{1}{8}} \left\| x^2 F_{\mu,\kappa}(t) \right\|_{\dot{H}^2} \Bigg].$$

2.4 Useful Estimates

In this section, we gather several lemmas that are used throughout the paper. First we introduce the decay properties for the (MKG).

Proposition 2.1 (Time Decay Effects for MKG) *Let $\theta, \kappa \in \{+, -\}$. Assume that (ϕ, A_μ) satisfies a priori assumption* (2.11)*, for given ε_1 and T. Then there exists C satisfying that for some $k := k(n) > 0$,*

$$\begin{aligned} \|\phi_\theta(t)\|_{W^{k,\infty}} &\leq C \langle t\rangle^{-2+\delta} \varepsilon_1, \\ \|A_{\mu,\kappa}(t)\|_{W^{k,\infty}} &\leq C \langle t\rangle^{-\frac{3}{2}} \varepsilon_1, \end{aligned} \tag{2.12}$$

for any $t \in [0, T]$, where δ is in (2.11)*.*

This proposition follows from the proof in [24]. This decay property (2.12) directly implies (1.5).

Lemma 2.2 (Coifman-Meyer Operator Estimates) *Let $1 \leq p, q \leq \infty$ satisfy that $\frac{1}{p} + \frac{1}{q} = \frac{1}{2}$. Assume that*

$$\|\mathbf{m}\|_{\mathrm{CM}} := \left\| \iint_{\mathbb{R}^{4+4}} \mathbf{m}(\xi, \zeta) e^{ix\cdot\xi} e^{iy\cdot\eta} d\xi d\eta \right\|_{L^1_{x,y}} \leq C_{\mathbf{m}} < \infty \quad (\zeta = \eta \ \ or \ \ \xi - \eta).$$

Then

$$\left\| \int_{\mathbb{R}^4} \mathbf{m}(\xi, \eta) \widehat{v}(\eta) \widehat{w}(\xi - \eta) d\eta \right\|_{L^2_\xi} \lesssim C_{\mathbf{m}} \|v\|_{L^p} \|w\|_{L^q}.$$

The following lemma is easily obtained.

Lemma 2.3 *Let* $u : \mathbb{R}^4 \to \mathbb{C}$*(or* $\mathbb{R}$*) and* $\mathcal{R}_j = \frac{\partial_j}{\langle D\rangle}$ *or* $\frac{\partial_j}{|D|}$*. Then we have*

$$\left\|\mathcal{R}_j u\right\|_{L^p} \lesssim \|u\|_{L^p},$$

for $j = 1, \cdots, 4$ *and* $1 < p < \infty$.

Lemma 2.4 *Let* $\theta, \kappa \in \{+, -\}$*. Assume that* ϕ, $A_\mu \in C([0, T], H^n)$ *satisfy a priori assumption* (2.11)*. Then*

$$\|\partial_t f_\theta(t)\|_{H^k} \lesssim \langle t\rangle^{-2+\frac{3\delta}{2}}\,\varepsilon_1^2, \tag{2.13}$$

$$\|\partial_t F_{\mu,\kappa}(t)\|_{H^k} \lesssim \langle t\rangle^{-1+\frac{\delta}{2}}\,\varepsilon_1^2, \tag{2.14}$$

where f_θ, $F_{\mu,\kappa}$ *are defined in* (2.4) *and k is in Proposition 2.1.*

Proof We first prove (2.13). Using (2.5), we see that

$$f_\theta(t) = \phi_{0,\theta} - \theta\frac{i}{2}\int_0^t e^{\theta is\langle D\rangle}\,\langle D\rangle^{-1}\left(\mathfrak{N}_1^{kg} + \mathfrak{N}_2^{kg} + \mathfrak{N}_3^{kg}\right)(s)\,ds.$$

Then we have

$$\partial_t f_\theta(t) = -\theta\frac{i}{2}e^{\theta it\langle D\rangle}\,\langle D\rangle^{-1}\left(\mathfrak{N}_1^{kg} + \mathfrak{N}_2^{kg} + \mathfrak{N}_3^{kg}\right)(t).$$

This yields that

$$\begin{aligned}
\|\partial_t f_\theta(t)\|_{H^k} &\lesssim \|A_{\mu,\theta_2}(t)\|_{H^k}\left(\|\partial_\mu\phi_{\theta_1}(t)\|_{W^{k-1,\infty}} + \|\phi_{\theta_1}(t)\|_{W^{k,\infty}}\right)\\
&\quad + \|A_{\mu,\theta_2}(t)\|_{H^k}\|\phi_{\theta_1}(t)\|_{W^{k,\infty}}^2\\
&\lesssim \langle t\rangle^{-2+\frac{3\delta}{2}}\,\varepsilon_1^2.
\end{aligned}$$

Let us move on to the proof of (2.14). Similarly, by (2.6), direct calculation leads us that

$$\partial_t F_{\mu,\kappa}(t) = -\kappa\frac{i}{2}e^{-\kappa it|D|}|D|^{-1}(\mathfrak{N}_1^m + \mathfrak{N}_2^m)(t).$$

By Sobolev embedding $L^{\frac{4}{3}} \hookrightarrow H^{-1}$ we have

$$\begin{aligned}
\left\|\partial_t F_{\mu,\kappa}(t)\right\|_{H^k} &\lesssim \left\|\phi_{\kappa_1}(t)\right\|_{L^4}\left\|\phi_{\kappa_2}(t)\right\|_{H^k} + \|A_\mu(s)\|_{H^k}\left\|\phi_{\kappa_1}(t)\right\|_{L^4}\left\|\phi_{\kappa_1}(t)\right\|_{W^{k,\infty}}\\
&\lesssim \langle t\rangle^{-1+\frac{\delta}{2}}\,\varepsilon_1^2,
\end{aligned}$$

and the desired conclusion follows. □

Lemma 2.5 *Let* $\theta, \kappa \in \{+, -\}$. *Assume that* $\phi, A_\mu \in C([0, T], H^n)$ *satisfy a priori assumption* (2.11). *Then*

$$\|x\partial_t f_\theta(t)\|_{H^k} \lesssim \langle t\rangle^{-\frac{1}{2}} \varepsilon_1^2.$$

Proof Since

$$\widehat{\partial_t f_\theta}(t,\xi) = -\theta\frac{i}{2}e^{\theta it\langle\xi\rangle}\langle\xi\rangle^{-1}\left(\widehat{\mathfrak{N}_1^{kg}}(t,\xi) + \widehat{\mathfrak{N}_2^{kg}}(t,\xi) + \widehat{\mathfrak{N}_3^{kg}}(t,\xi)\right),$$

we obtain

$$\nabla_\xi\widehat{\partial_t f_\theta}(t,\xi) = -\theta\frac{i}{2}e^{\theta it\langle\xi\rangle}\left(\theta it\frac{\xi}{\langle\xi\rangle^2} - \frac{\xi}{\langle\xi\rangle^3}\right)\left(\widehat{\mathfrak{N}_1^{kg}}(t,\xi) + \widehat{\mathfrak{N}_2^{kg}}(t,\xi) + \widehat{\mathfrak{N}_3^{kg}}(t,\xi)\right) \tag{2.15}$$

$$-\theta\frac{i}{2}e^{\theta it\langle\xi\rangle}\langle\xi\rangle^{-1}\nabla_\xi\left(\widehat{\mathfrak{N}_1^{kg}}(t,\xi) + \widehat{\mathfrak{N}_2^{kg}}(t,\xi) + \widehat{\mathfrak{N}_3^{kg}}(t,\xi)\right). \tag{2.16}$$

Then we have

$$\|(2.15)\|_{H^k} \lesssim t\sum_{j=1}^{3}\|\mathfrak{N}_j^{kg}(t)\|_{H^{k-1}} \lesssim \langle t\rangle^{-\frac{1}{2}}\varepsilon_1^2.$$

To handle (2.16), we have

$$x\phi_\theta(t) = xe^{\theta it\langle D\rangle}f_\theta(t) = \theta it\langle D\rangle\phi_\theta(t) + e^{\theta it\langle D\rangle}xf_\theta(t),$$

which implies that

$$\|x\phi_\theta(t)\|_{H^k} \lesssim \langle t\rangle\varepsilon_1.$$

This leads us that, for $j = 1, 2, 3$,

$$\|x\mathfrak{N}_j^{kg}(t)\|_{H^{k-1}} \lesssim \langle t\rangle^{-\frac{1}{2}}\varepsilon_1^2.$$

This completes the proof. □

3 Proof of the Main Theorem

Given any $T > 0$, let ϕ, A_μ be solutions with initial data satisfying (1.4) on $[0, T]$. For some small $0 < \varepsilon_1 < 1$ and $\theta, \kappa \in \{+, -\}$ we assume that

$$\|\phi_\theta\|_{\Sigma^{\mathbf{D}}_{T,\theta}}, \|A_{\mu,\kappa}\|_{\Sigma^{\mathbf{M}}_{T,\kappa}} \leq K\varepsilon_1.$$

for some absolute constant $K > 0$. Then there exists C depending only on K such that

$$\|\phi_\theta\|_{\Sigma^{\mathbf{D}}_{T,\theta}}, \|A_{\mu,\kappa}\|_{\Sigma^{\mathbf{M}}_{T,\kappa}} \leq \varepsilon_0 + C\varepsilon_1^2. \tag{3.1}$$

From (3.1), Proposition 2.1 exhibit the global bound (1.5). The claim (3.1) will turn out to follow from Proposition 3.1 below. Then the a priori estimate with a bootstrap argument yields the global existence for sufficiently small ε_0. Moreover, by the time reversibility, we obtain a solution for all times.

Now we introduce the following energy estimates.

Proposition 3.1 *Let $\theta, \kappa \in \{+, -\}$. Assume that (ϕ, A_μ) satisfy the a priori bound* (2.11) *for given ε_1 and T. Let $(\phi_\theta, A_{\mu,\kappa}) \in C([0, T], H^n)$ be a solution to* (2.1) *with initial data satisfying* (1.4). *Then we obtain the following estimates:*

$$\sup_{t\in[0,T]} \|\phi_\theta(t)\|_{H^n} \leq \varepsilon_0 + C\varepsilon_1^2, \tag{3.2}$$

$$\sup_{t\in[0,T]} \|xf_\theta(t)\|_{L^2} \leq \varepsilon_0 + C\varepsilon_1^2, \tag{3.3}$$

$$\sup_{t\in[0,T]} \langle t\rangle^{-\delta+\beta} \|x^2 f_\theta(t)\|_{L^2} \leq \varepsilon_0 + C\varepsilon_1^2. \tag{3.4}$$

If $1 \leq \ell \leq n$, then

$$\sup_{t\in[0,T]} \left[\langle t\rangle^{-\frac{\delta}{2}} \|A_{\mu,\kappa}(t)\|_{H^n} + \|A_{\mu,\kappa}(t)\|_{\dot{H}^\ell}\right] \leq \varepsilon_0 + C\varepsilon_1^2, \tag{3.5}$$

$$\sup_{t\in[0,T]} \langle t\rangle^{-\frac{\delta}{2}} \|xF_{\mu,\kappa}(t)\|_{\dot{H}^1} \leq \varepsilon_0 + C\varepsilon_1^2, \tag{3.6}$$

$$\sup_{t\in[0,T]} \langle t\rangle^{-\frac{1}{8}} \|x^2 F_{\mu,\kappa}(t)\|_{\dot{H}^2} \leq \varepsilon_0 + C\varepsilon_1^2. \tag{3.7}$$

3.1 Estimates for High Sobolev Part

We begin with the estimates of (3.2) and (3.5). In fact, the energy estimates (3.2) and (3.5) can be readily shown as follows: We first consider $\mathfrak{N}_1^{kg}$. To this end, it suffices to show

$$\left\| \int_0^t e^{-\theta_0 is\langle D\rangle} \langle D\rangle^{-1} \mathfrak{N}_1^{kg}(s)ds \right\|_{H^n} \lesssim \varepsilon_1^2. \tag{3.8}$$

By Lorenz gauge $\partial^\mu A_\mu = 0$ and integration by parts in time, we see that

$$\begin{aligned}
&\int_0^t e^{-\theta_0 is\langle D\rangle} \langle D\rangle^{-1} \mathfrak{N}_1^{kg}(s)ds \\
&\quad = e^{-\theta_0 is\langle D\rangle} \langle D\rangle^{-1} (A_0(s)\phi(s)) \Big|_{s=t,0} \\
&\quad\quad + \int_0^t e^{-\theta_0 is\langle D\rangle} \Big(\langle D\rangle^{-1} [\partial_s A_0(s)\phi(s) + A^j(s)\partial_j\phi(s)] - \theta_0 i A_0(s)\phi(s) \Big) ds.
\end{aligned}$$

Then we estimate for $k = 0, 1$,

$$\begin{aligned}
\|A_0(t)\phi(t)\|_{H^n} &\lesssim \|A_0(t)\|_{H^n} \|\phi(t)\|_{L^\infty} + \|A_0(t)\|_{L^\infty} \|\phi(t)\|_{H^n} \\
&\lesssim \varepsilon_1^2 \langle t\rangle^{-\frac{3}{2}}, \\
\| \langle D\rangle^{-1} \left(\partial_j^{1-k} A_j(s)\partial_j^k\phi(s) \right) \|_{H^n} &\lesssim \left\|A_j(s)\right\|_{H^n} \|\phi(s)\|_{W^{1,\infty}} + \left\|A_j(s)\right\|_{W^{1,\infty}} \|\phi(s)\|_{H^n} \\
&\lesssim \varepsilon_1^2 \langle s\rangle^{-\frac{3}{2}},
\end{aligned}$$

which imply (3.8). Similarly, we may get the desired bound for $\mathfrak{N}_2^{kg}$. For $\mathfrak{N}_3^{kg}$, we estimate straightforwardly

$$\begin{aligned}
\| \langle D\rangle^{-1} \mathfrak{N}_3^{kg}(s)\|_{H^n} &\lesssim \|A^\mu(s)\|_{H^n} \|A_\mu(s)\|_{L^\infty} \|\phi(s)\|_{L^\infty} \\
&\quad + \|A^\mu(s)\|_{L^\infty} \|A_\mu(s)\|_{L^\infty} \|\phi(s)\|_{H^n} \lesssim \langle s\rangle^{-\frac{3}{2}}.
\end{aligned}$$

Then, these nonlinear estimates yield that

$$\|\phi_\theta(t)\|_{H^n} \le \varepsilon_0 + C\varepsilon_1^2.$$

Similarly to the estimates for $\mathfrak{N}_2^{kg}$ and $\mathfrak{N}_3^{kg}$, if $\ell \ge 1$, then we get

$$\|A_{\mu,\kappa}(t)\|_{\dot{H}^\ell} \le \varepsilon_0 + C\varepsilon_1^2 \tag{3.9}$$

and if $\ell = 0$, then, by Sobolev embedding $L^{\frac{4}{3}} \hookrightarrow H^{-1}$ and charge conservation law, we have

$$\begin{aligned}
\|A_{\mu,\kappa}(t)\|_{L^2} &\leq \|a_{\mu,\kappa}\|_{L^2} + C\int_0^t \|\langle\phi(s), \alpha_\mu\phi(s)\rangle\|_{L^{\frac{4}{3}}}\, ds \\
&\leq \varepsilon_0 + C\int_0^t \|\phi(s)\|_{L^2}^{\frac{3}{2}} \|\phi(s)\|_{L^\infty}^{\frac{1}{2}}\, ds \\
&\leq \varepsilon_0 + C\varepsilon_1^2 \int_0^t \langle s\rangle^{-1+\frac{\delta}{2}}\, ds \\
&\leq \varepsilon_0 + C\,\langle t\rangle^{\frac{\delta}{2}}\,\varepsilon_1^2.
\end{aligned} \tag{3.10}$$

From (3.9) and (3.10), we see that (3.5).

3.2 *Scattering Results for Maxwell-Klein-Gordon System*

In this section, we focus on the scattering phenomena of Klein-Gordon spinor ϕ and potential A_μ. Given the decay properties for Klein-Gordon and Maxwell parts, the proof of scattering follows straightforwardly. For the sake of convenience of readers, we outline the proof below. For $t_1 \leq t_2 \in [0, T]$, we have

$$\begin{aligned}
\|f_\theta(t_2) - f_\theta(t_1)\|_{L^2} &\lesssim \left\| \int_{t_1}^{t_2} e^{-\theta is\langle D\rangle} A_\mu(s)\alpha^\mu\phi(s)ds \right\|_{L^2} \\
&\lesssim \int_{t_1}^{t_2} \left\|A_\mu(s)\alpha^\mu\phi(s)\right\|_{L^2} ds \lesssim \langle t_1\rangle^{-\frac{1}{2}}\,\varepsilon_1^2.
\end{aligned}$$

We define $\phi_\theta^\infty = \lim_{t\to\infty} f_\theta(t)$, where the limit is taken in L^2. Then we obtain

$$\left\|\phi_\theta(t) - e^{\theta it\langle D\rangle}\phi_\theta^\infty\right\|_{L^2} \xrightarrow{t\to\infty} 0.$$

By putting $\phi^\infty(t) = e^{-it\langle D\rangle}\phi_+^\infty + e^{it\langle D\rangle}\phi_-^\infty$, this proves the scattering phenomenon for Klein-Gordon spinor.

Analogously, we see that, for $t_1 \leq t_2 \in [0, T]$,

$$\left\|F_{\mu,\theta}(t_2) - F_{\mu,\theta}(t_1)\right\|_{\dot{H}^1} \lesssim \langle t_1\rangle^{-1+\delta}\,\varepsilon_1^2,$$

which implies that

$$\left\|A_{\mu,\theta}(t) - e^{\theta it|D|}B_{\mu,\theta}^\infty\right\|_{\dot{H}^1} \xrightarrow{t\to\infty} 0,$$

by setting $B_{\mu,\theta}^{\infty} := \lim_{t\to\infty} F_{\mu,\theta}(t)$ in $\dot{H}^1$. Defining $A_\mu^{\infty}(t) := e^{it|D|}B_{\mu,\theta}^{\infty} + e^{-it|D|}B_{\mu,\theta}^{\infty}$, we estimate

$$\left\|A_\mu(t) - A_\mu^{\infty}(t)\right\|_{\dot{H}^1} + \left\|\partial_t A_\mu(t) - \partial_t A_\mu^{\infty}(t)\right\|_{L^2} \xrightarrow{t\to\infty} 0.$$

4 Weighted Energy Estimates for Scalar Field: Proof of (3.3) and (3.4)

4.1 *Proof of (3.3)*

In order to estimate xf_θ in L^2, we use Plancherel's theorem (2.7) as follows:

$$\|xf_\theta(t)\|_{L^2} = \left\|\nabla_\xi \widehat{f_\theta}(t,\xi)\right\|_{L^2_\xi}.$$

This implies that, for $\theta_0 \in \{+,-\}$,

$$\nabla_\xi \widehat{f_{\theta_0}}(t,\xi) = \sum_{\theta_1,\theta_2\in\{\pm\}} \int_0^t \nabla_\xi \int_{\mathbb{R}^4} e^{isp_\Theta(\xi,\eta)} \frac{\xi^j}{\langle\xi\rangle} \widehat{F_{\mu,\theta_2}}(s,\eta)\widehat{f_{\theta_1}}(s,\xi-\eta)d\eta ds \tag{4.1}$$

$$+ \sum_{\theta_1,\theta_2\in\{\pm\}} \int_0^t \nabla_\xi \int_{\mathbb{R}^4} e^{isp_\Theta(\xi,\eta)} \frac{\xi_j - \eta_j}{\langle\xi\rangle} \widehat{F_{\mu,\theta_2}}(s,\eta)\widehat{f_{\theta_1}}(s,\xi-\eta)d\eta ds \tag{4.2}$$

$$+ \sum_{\theta_1,\theta_2\in\{\pm\}} \int_0^t \nabla_\xi \partial_s \int_{\mathbb{R}^4} e^{isp_\Theta(\xi,\eta)} \frac{1}{\langle\xi\rangle} \widehat{F_{\mu,\theta_2}}(s,\eta)\widehat{f_{\theta_1}}(s,\xi-\eta)d\eta ds \tag{4.3}$$

$$+ \sum_{\theta_1,\theta_2\in\{\pm\}} \int_0^t \nabla_\xi \int_{\mathbb{R}^4} e^{isp_\Theta(\xi,\eta)} \frac{1}{\langle\xi\rangle} \widehat{F_{\mu,\theta_2}}(s,\eta)\partial_s\widehat{f_{\theta_1}}(s,\xi-\eta)d\eta ds \tag{4.4}$$

$$+ \int_0^t \nabla_\xi \int_{\mathbb{R}^4} e^{-\theta_0 is\langle\xi\rangle} \frac{1}{\langle\xi\rangle} \widehat{A^\mu A_\mu}(s,\eta)\widehat{\phi}(s,\xi-\eta)d\eta ds, \tag{4.5}$$

where p_Θ is defined in (2.9). Note that (4.1) and (4.2) come from the spatial derivative in the nonlinearity of Klein-Gordon part. On the other hand, (4.3) and (4.4) also come from the time derivative. The estimates for (4.1) and (4.2) are almost the same as those in Section 6 of [24] from the fact that Riesz transform is negligible in our analysis. Hence, we only focus on the estimates for (4.3)–(4.5) in this section.

Estimates for (4.3) This estimate can be divided into the cases that the time derivative falls on the phase $e^{isp_\Theta(\xi,\eta)}$, spinor $f_{\theta_1}(s)$, and gauge fields $F_{\mu,\theta_2}(s)$. We begin with the estimates for the first case. Then, we have

$$\int_0^t \nabla_\xi \int_{\mathbb{R}^4} e^{isp_\Theta(\xi,\eta)} \frac{p_\Theta(\xi,\eta)}{\langle\xi\rangle} \widehat{F_{\mu,\theta_2}}(s,\eta)\widehat{f_{\theta_1}}(s,\xi-\eta)d\eta ds.$$

Since the multiplier $\frac{p_\Theta(\xi,\eta)}{\langle\xi\rangle}$ is negligible in terms of Lemma 2.2, this case can be done similarly to the estimates for (4.1) or (4.2). Let us move on to the case that time derivative falls on f_{θ_1} or F_{μ,θ_2}. Then, it follows from Lemma 2.4 more easily. Indeed, if the ∇_ξ falls on the phase, by Lemma 2.4, we have

$$\begin{aligned}
\|(4.3)\|_{L^2_\xi} &\lesssim \int_0^t s \left\| \int_{\mathbb{R}^4} e^{isp_\Theta(\xi,\eta)} \frac{\nabla_\xi p_\Theta(\xi,\eta)}{\langle\xi\rangle} \partial_s \left[\widehat{F_{\mu,\theta_2}}(s,\eta)\widehat{f_{\theta_1}}(s,\xi-\eta)\right] d\eta \right\|_{L^2_\xi} ds \\
&\lesssim \int_0^t s \left(\|\partial_s F_{\mu,\theta_2}(s)\|_{L^2} \|\phi_{\theta_1}(s)\|_{L^\infty} + \|A_{\mu,\theta_2}(s)\|_{L^\infty} \|\partial_s f_{\theta_1}(s)\|_{L^2} \right) ds \\
&\lesssim \int_0^t \varepsilon_1^3 s \langle s\rangle^{-3+\frac{3\delta}{2}} ds.
\end{aligned}$$

If the derivative ∇_ξ falls on $\widehat{f_{\theta_1}}$, Lemma 2.5 completes the proof.

Estimates for (4.4) This case can be done by the above estimates when the time derivative falls on f_{θ_1}.

Estimates for (4.5) When the derivative falls on the phase function $e^{\theta_0 is\langle\xi\rangle}$, we have

$$\|(4.5)\|_{L^2_\xi} \lesssim \int_0^t s\|A^\mu(s)\|_{L^2}\|A_\mu(s)\|_{L^\infty}\|\phi(s)\|_{L^\infty} ds \lesssim \int_0^t \varepsilon_1^3 \langle s\rangle^{-\frac{5}{2}+\frac{3\delta}{2}} ds \lesssim \varepsilon_1^3.$$

Then it suffices to show

$$\int_0^t \left\| \langle D\rangle^{-1} x \left(A^\mu A_\mu \phi\right)(s) \right\|_{L^2} ds \lesssim \varepsilon_1^3.$$

Since

$$xA_{\mu,\theta}(t) = \theta it \frac{\nabla}{|D|} A_{\mu,\theta}(t) + e^{\theta it|D|} x F_{\mu,\theta}(t),$$

we have

$$\begin{aligned}
\|xA_\mu(s)\|_{L^4} &\lesssim \sum_{\theta\in\{\pm\}} \|xA_{\mu,\theta}(s)\|_{L^4} \lesssim \sum_{\theta\in\{\pm\}} \left(s\|A_{\mu,\theta}(s)\|_{L^4} + \|xF_{\mu,\theta}(s)\|_{\dot H^1} \right) \\
&\lesssim \varepsilon_1 \langle s\rangle^{\frac{1}{4}+\frac{\delta}{2}}, \qquad (4.6)
\end{aligned}$$

which implies that

$$\int_0^t \left\| \langle D\rangle^{-1} x \left(A^\mu A_\mu \phi\right)(s)\right\|_{L^2} ds \lesssim \int_0^t \|xA^\mu(s)\|_{L^4}\|A_\mu(s)\|_{L^4}\|\phi(s)\|_{L^\infty} ds$$
$$\lesssim \int_0^t \varepsilon_1^3 \langle s\rangle^{-\frac{5}{2}+2\delta} ds.$$

4.2 Proof of (3.4)

In this section, we devote to prove (3.4). We begin with

$$\left\|x^2 f_\theta(t)\right\|_{L^2} = \left\|\nabla_\xi^2 \widehat{f_\theta}(t,\xi)\right\|_{L^2_\xi},$$

by Plancherel's theorem. Given the estimates for (4.1)–(4.5), the cases that ∇_ξ^2 falls on $\mathfrak{N}_1^{kg}$ and $\mathfrak{N}_2^{kg}$ can be done in an identical way to the estimates detailed in Section 6 of [24]. Therefore, it suffices to consider only the following case:

$$\int_0^t \nabla_\xi^2 \int_{\mathbb{R}^4} e^{-\theta_0 is\langle\xi\rangle} \frac{1}{\langle\xi\rangle} \widehat{A^\mu A_\mu}(s,\eta)\widehat{\phi}(s,\xi-\eta)d\eta ds. \tag{4.7}$$

If all derivatives fall on only the phase function $e^{\theta_0 is\langle\xi\rangle}$, we estimate

$$\|(4.7)\|_{L^2_\xi} \lesssim \int_0^t s^2\|A^\mu(s)\|_{H^1}\|A_\mu(s)\|_{W^{1,\infty}}\|\phi(s)\|_{W^{1,\infty}} ds \lesssim \int_0^t \varepsilon_1^3 \langle s\rangle^{-\frac{3}{2}+\frac{3\delta}{2}} ds$$
$$\lesssim \varepsilon_1^3.$$

If the one derivative falls on $e^{\theta_0 is\langle\xi\rangle}$ and the other one falls on $A^\mu A_\mu \phi$, we have

$$\|(4.7)\|_{L^2_\xi} \lesssim \int_0^t s\|xA^\mu(s)\|_{L^4}\|A_\mu(s)\|_{L^4}\|\phi(s)\|_{L^\infty} ds \lesssim \int_0^t \varepsilon_1^3 \langle s\rangle^{-\frac{3}{2}+2\delta} ds \lesssim \varepsilon_1^3.$$

The case that the all derivatives fall on $A^\mu A_\mu \phi$ can be treated as follows:

$$\|(4.7)\|_{L^2_\xi} \lesssim \int_0^t \left\|\langle D\rangle^{-1} x^2(A^\mu A_\mu \phi)(s)\right\|_{L^2} ds \lesssim \int_0^t \|xA^\mu(s)\|_{L^4}\|xA_\mu(s)\|_{L^4}\|\phi(s)\|_{L^\infty} ds$$
$$\lesssim \int_0^t \varepsilon_1^3 \langle s\rangle^{-\frac{3}{2}+2\delta} ds \lesssim \varepsilon_1^3.$$

5 The Weighted Energy Estimates for Gauge: Proof of (3.6) and (3.7)

5.1 *Proof of (3.6)*

By Plancherel's theorem, one gets that

$$\left\| x F_{\mu,\kappa}(t) \right\|_{\dot{H}^1} = \left\| |\xi| \nabla_\xi \widehat{F_{\mu,\kappa}}(\xi) \right\|_{L^2_\xi}.$$

Recalling the notation (2.8), (2.10), and Duhamel's formula (2.3), $|\xi|\nabla_\xi F_{\mu,\kappa_0}$ can be represented by the following:

$$\int_0^t |\xi|\nabla_\xi \int_{\mathbb{R}^4} e^{isq_K(\xi,\eta)} |\xi|^{-1} \widehat{f_{\kappa_1}}(s,\xi+\eta)\eta_j \overline{\widehat{f_{\kappa_2}}(s,\eta)}\, d\eta ds, \tag{5.1}$$

$$\int_0^t |\xi|\nabla_\xi \int_{\mathbb{R}^4} e^{isq_K(\xi,\eta)} |\xi|^{-1} \widehat{f_{\kappa_1}}(s,\xi+\eta)\overline{\partial_s \widehat{f_{\kappa_2}}(s,\eta)}\, d\eta ds, \tag{5.2}$$

$$\int_0^t |\xi|\nabla_\xi \int_{\mathbb{R}^4} e^{-\kappa_0 is|\xi|} |\xi|^{-1} \widehat{A_\mu}(s,\xi-\eta)\widehat{|\phi|^2}(s,\eta)\, d\eta ds, \tag{5.3}$$

where q_K is defined in (2.10). The estimates for (5.1) can be handled in an identical way to Section 7 of [24]. By (2.13), we also have the desired estimates for (5.2) straightforwardly. In (5.3), we consider the following terms:

$$\int_0^t \int_{\mathbb{R}^4} e^{-\kappa_0 is|\xi|} \left(-\kappa_0 is\frac{\xi}{|\xi|} + \frac{\xi}{|\xi|^2} \right) \widehat{A_\mu}(s,\xi-\eta)\widehat{|\phi|^2}(s,\eta)\, d\eta ds, \tag{5.4}$$

$$\int_0^t \int_{\mathbb{R}^4} e^{-\kappa_0 is|\xi|} \widehat{xA_\mu}(s,\xi-\eta)\widehat{|\phi|^2}(s,\eta)\, d\eta ds. \tag{5.5}$$

Since $L^{\frac{3}{4}} \hookrightarrow H^{-1}$, we have

$$\begin{aligned}\|(5.4)\|_{L^2_\xi} &\lesssim \int_0^t \left(s\|A_\mu(s)\|_{L^2}\|\phi(s)\|_{L^\infty} + \|A_\mu(s)\|_{L^2}\|\phi(s)\|_{L^4} \right) \|\phi(s)\|_{L^\infty} ds \\ &\lesssim \int_0^t \varepsilon_1^3 \langle s\rangle^{-2+\frac{5\delta}{2}}\, ds \lesssim \varepsilon_1^3.\end{aligned}$$

Using (4.6), we see that

$$\|(5.5)\|_{L^2_\xi} \lesssim \int_0^t \|xA_\mu(s)\|_{L^4}\|\phi(s)\|_{L^8}^2 ds \lesssim \int_0^t \varepsilon_1^3 \langle s\rangle^{-\frac{11}{4}+2\delta}\, ds \lesssim \varepsilon_1^3.$$

5.2 Proof of (3.7)

We begin with

$$\left\| x^2 F_{\mu,\kappa}(t) \right\|_{\dot{H}^2} = \left\| |\xi|^2 \nabla_\xi^2 \widehat{F_{\mu,\kappa}}(t,\xi) \right\|_{L^2_\xi}.$$

Similarly to Sect. 5.1, it suffices to handle the following terms:

$$\int_0^t |\xi| \nabla_\xi \int_{\mathbb{R}^4} e^{isq_\kappa(\xi,\eta)} |\xi|^{-1} \widehat{f_{\kappa_1}}(s,\xi+\eta) \eta_j \overline{\widehat{f_{\kappa_2}}(s,\eta)}\, d\eta ds, \tag{5.6}$$

$$\int_0^t |\xi|^2 \nabla_\xi^2 \int_{\mathbb{R}^4} e^{isq_\kappa(\xi,\eta)} |\xi|^{-1} \widehat{f_{\kappa_1}}(s,\xi+\eta) \overline{\partial_s \widehat{f_{\kappa_2}}(s,\eta)}\, d\eta ds, \tag{5.7}$$

$$\int_0^t |\xi|^2 \nabla_\xi^2 \int_{\mathbb{R}^4} e^{-\kappa_0 is|\xi|} |\xi|^{-1} \widehat{A_\mu}(s,\xi-\eta) \widehat{|\phi|^2}(s,\eta)\, d\eta ds, \tag{5.8}$$

and the estimates for (5.6) and (5.7) can be treated in an identical way to those in Section 8 of [24]. We omit the details. We focus on the estimates for (5.8). This case can be divided into the following cases:

$$\int_0^t s^2 \int_{\mathbb{R}^4} e^{-\kappa_0 is|\xi|} \frac{\xi^2}{|\xi|} \widehat{A_\mu}(s,\xi-\eta) \widehat{|\phi|^2}(s,\eta)\, d\eta ds, \tag{5.9}$$

$$\int_0^t s \int_{\mathbb{R}^4} e^{-\kappa_0 is|\xi|} \left(\frac{\xi^2}{|\xi|^2} + \nabla_\xi \cdot \xi \right) \widehat{A_\mu}(s,\xi-\eta) \widehat{|\phi|^2}(s,\eta)\, d\eta ds, \tag{5.10}$$

$$\int_0^t s \int_{\mathbb{R}^4} e^{-\kappa_0 is|\xi|} \xi \widehat{xA_\mu}(s,\xi-\eta) \widehat{|\phi|^2}(s,\eta)\, d\eta ds, \tag{5.11}$$

$$\int_0^t \int_{\mathbb{R}^4} e^{-\kappa_0 is|\xi|} |\xi| \widehat{x^2 A_\mu}(s,\xi-\eta) \widehat{|\phi|^2}(s,\eta)\, d\eta ds. \tag{5.12}$$

By Hölder inequality, we obtain

$$\|(5.9)\|_{L^2_\xi} + \|(5.10)\|_{L^2_\xi} \lesssim \int_0^t s^2 \|A_\mu(s)\|_{H^1} \|\phi(s)\|^2_{W^{1,\infty}} ds \lesssim \int_0^t \varepsilon_1^3 \langle s \rangle^{-2+\frac{5\delta}{2}}\, ds \lesssim \varepsilon_1^3.$$

It also follows from (4.6) that

$$\|(5.11)\|_{L^2_\xi} \lesssim \int_0^t s \|xA_\mu(s)\|_{L^4} \|\phi(s)\|^2_{L^8} ds \lesssim \int_0^t \varepsilon_1^3 \langle s \rangle^{-\frac{7}{4}+2\delta}\, ds \lesssim \varepsilon_1^3.$$

For (5.12),

$$\|(5.12)\|_{L^2_\xi} \lesssim \int_0^t \left\| x^2 A_\mu(s)|\phi(s)|^2 \right\|_{\dot{H}^1} ds \lesssim \int_0^t \left\| x^2 A_\mu(s)|\phi(s)|^2 \right\|_{\dot{H}^1} ds.$$

Since

$$x\phi_\theta(s) = \theta is \frac{\nabla}{\langle D \rangle}\phi_\theta(s) + e^{\theta is\langle D\rangle} x f_\theta(s)$$

by Lemma 2.3, we have

$$\|x\phi(s)\|_{L^4} \lesssim \sum_{\theta\in\{\pm\}} \|x\phi_\theta(s)\|_{L^4} \lesssim \sum_{\theta\in\{\pm\}} \left(s\|\phi_\theta(s)\|_{L^4} + \|xf_\theta(s)\|_{H^1}\right) \lesssim \varepsilon_1 \langle s\rangle^\delta . \tag{5.13}$$

This leads us that

$$\begin{aligned}\|(5.12)\|_{L^2_\xi} &\lesssim \int_0^t \|xA_\mu(s)\|_{W^{1,4}} \|x\phi(s)\|_{W^{1,4}} \|\phi(s)\|_{W^{1,\infty}} ds \lesssim \int_0^t \varepsilon_1^3 \langle s\rangle^{-\frac{7}{4}+\frac{5\delta}{2}} ds \\ &\lesssim \varepsilon_1^3.\end{aligned}$$

Note that we used $W^{1,p}$-spaces in (4.6) and (5.13) instead of L^p-spaces.

Acknowledgments I would like to express my gratitude to Professor Vladimir Simeonov Georgiev and Tohru Ozawa for organizing the Minisymposia at ICIAM 2023 Tokyo and this proceeding. I also thank the anonymous referees for valuable suggestions.

The author was supported in part by RS-2025-00514043 and RS-2024-00463260, the National Research Foundation of Korea (NRF) grant funded by the Korea government (MSIT) and (MOE), respectively.

References

1. Candy, T., Herr, S.: Transference of bilinear restriction estimates to quadratic variation norms and the Dirac-Klein-Gordon system. Anal. PDE **11**(5), 1171–1240 (2018)
2. Candy, T., Kauffman, C., Lindblad, H.: Asymptotic behavior of the Maxwell-Klein-Gordon system. Commun. Math. Phys. **367**(2), 683–716 (2019)
3. Cho, Y., Lee, K.: Global dynamics for the Maxwell-Dirac system. Preprint. arXiv.org/abs/2406.18887 (2024)
4. Dong, S., Wyatt, Z.: Asymptotic stability for the Dirac-Klein-Gordon system in two space dimensions. Ann. Inst. H. Poincaré C Anal. Non Linéaire. **41**(6), 1419–1464 (2024)
5. Dong, S., Li, K., Ma, Y., Yuan, X.: Global behavior of small data solutions for the 2D Dirac-Klein-Gordon system. Trans. Amer. Math. Soc. **377**(1), 649–695 (2024)
6. Eardley, D.M., Moncrief, V.: The global existence of Yang-Mills-Higgs fields in 4-dimensional Minkowski space. I. Local existence and smoothness properties. Commun. Math. Phys. **83**(2), 171–191 (1982)

7. Eardley, D.M., Moncrief, V.: The global existence of Yang-Mills-Higgs fields in 4-dimensional Minkowski space. II. Completion of proof. Commun. Math. Phys. **83**(2), 193–212 (1982)
8. Fang, A., Wang, Q., Yang, S.: Global solution for massive Maxwell-Klein-Gordon equations with large Maxwell field. Ann. PDE **7**(1), Paper No. 3, 69 (2021)
9. Gavrus, C.: Global well-posedness for the massive Maxwell-Klein-Gordon equation with small critical Sobolev data. Ann. PDE **5**(1), Paper No. 10, 101 (2019)
10. Germain, P., Masmoudi, N., Shatah, J.: Global solutions for 3D quadratic Schrödinger equations. Int. Math. Res. Not. **2009**(3), 414–432 (2008)
11. Germain, P., Masmoudi, N., Shatah, J.: Global solutions for 2D quadratic Schrödinger equations. J. Math. Pures Appl. **97**(5), 505–543 (2012)
12. Germain, P., Masmoudi, N., Shatah, J.: Global solutions for the gravity water waves equation in dimension 3. Ann. Math. (2) **175**(2), 691–754 (2012)
13. Gustafson, S., Nakanishi, K., Tsai, T.P.: Scattering theory for the Gross-Pitaevskii equation in three dimensions. Commun. Contemp. Math. **11**(4), 657–707 (2009)
14. Herr, S., Ifrim, M., Spitz, M.: Modified scattering for the three dimensional Maxwell-Dirac system. Preprint, arXiv:2406.02460 (2024)
15. Ionescu, A.D., Pausader, B.: Global solutions of quasilinear systems of Klein-Gordon equations in 3D. J. Eur. Math. Soc. (JEMS) **16**(11), 2355–2431 (2014)
16. Ionescu, A.D., Pausader, B.: On the global regularity for a wave-Klein-Gordon coupled system. Acta Math. Sin. (Engl. Ser.) **35**(6), 933–986 (2019)
17. Ionescu, A.D., Pausader, B.: The Einstein-Klein-Gordon Coupled System: Global Stability of the Minkowski Solution, volume 213 of Annals of Mathematics Studies. Princeton University Press, Princeton (2022)
18. Klainerman, S., Machedon, M.: Space-time estimates for null forms and the local existence theorem. Comm. Pure Appl. Math. **46**(9), 1221–1268 (1993)
19. Klainerman, S., Machedon, M.: On the Maxwell-Klein-Gordon equation with finite energy. Duke Math. J. **74**(1), 19–44 (1994)
20. Klainerman, S., Machedon, M.: Finite energy solutions of the Yang-Mills equations in $\mathbf{R}^{3+1}$. Ann. Math. (2) **142**(1), 39–119 (1995)
21. Klainerman, S., Wang, Q., Yang, S.: Global solution for massive Maxwell-Klein-Gordon equations. Commun. Pure Appl. Math. **73**(1), 63–109 (2020)
22. Krieger, J., Lührmann, J.: Concentration compactness for the critical Maxwell-Klein-Gordon equation. Ann. PDE **1**(1), Art. 5, 208 (2015)
23. Krieger, J., Sterbenz, J., Tataru, D., Global well-posedness for the Maxwell-Klein-Gordon equation in $4+1$ dimensions: small energy. Duke Math. J. **164**(6), 973–1040 (2015)
24. Lee, K.: Scattering results for the (1+4) dimensional massive Maxwell-Dirac system under lorenz gauge condition. Preprint, arXiv:2312.13621 (2023)
25. Oh, S.J., Tataru, D.: Global well-posedness and scattering of the $(4+1)$-dimensional Maxwell-Klein-Gordon equation. Invent. Math. **205**(3), 781–877 (2016)
26. Oh, S.J., Tataru, D.: Energy dispersed solutions for the $(4+1)$-dimensional Maxwell-Klein-Gordon equation. Am. J. Math. **140**(1), 1–82 (2018)
27. Ouyang, Z.: Modified wave operators for the Wave-Klein-Gordon system. Adv. Math. **423**, Paper No. 109042, 84 (2023)
28. Pusateri, F.: Modified scattering for the Boson star equation. Commun. Math. Phys. **332**(3), 1203–1234 (2014)
29. Rodnianski, I., Tao, T.: Global regularity for the Maxwell-Klein-Gordon equation with small critical Sobolev norm in high dimensions. Commun. Math. Phys. **251**(2), 377–426 (2004)
30. Selberg, S., Tesfahun, A.: Finite-energy global well-posedness of the Maxwell-Klein-Gordon system in Lorenz gauge. Commun. Partial Differ. Equations **35**(6), 1029–1057 (2010)
31. Yang, S.: On the global behavior of solutions of the Maxwell-Klein-Gordon equations. Adv. Math. **326**, 490–520 (2018)
32. Yang, S., Yu, P.: On global dynamics of the Maxwell-Klein-Gordon equations. Camb. J. Math. **7**(4), 365–467 (2019)

A Short Review on Improvements and Stability for Some Interpolation Inequalities

Jean Dolbeault, Maria J. Esteban, Alessio Figalli, Rupert L. Frank, and Michael Loss

Abstract In this paper, we present recent stability results with explicit and dimensionally sharp constants and optimal norms for the Sobolev inequality and for the Gaussian logarithmic Sobolev inequality obtained by the authors in Dolbeault et al. (Sharp stability for Sobolev and log-Sobolev inequalities, with optimal dimensional dependence, Camb. J. Math., 2025). The stability for the Gaussian logarithmic Sobolev inequality was obtained as a byproduct of the stability for the Sobolev inequality. Here we give a new, direct, alternative proof. We also discuss improved versions of interpolation inequalities based on the *carré du champ* method.

Keywords Sobolev inequality · Logarithmic Sobolev inequality · Interpolation · Gagliardo-Nirenberg inequalities · Stability · Rearrangement · Steiner symmetrization · Heat equation · Fast diffusion equations

2020 Mathematics Subject Classification Primary: 49J40; Secondary: 26D10, 35A23, 58E35

J. Dolbeault · M. J. Esteban (✉)
CEREMADE (CNRS UMR No. 7534), PSL University, Université Paris-Dauphine, Paris, France
e-mail: dolbeaul@ceremade.dauphine.fr; esteban@ceremade.dauphine.fr

A. Figalli
Mathematics Department, ETH Zürich, Zürich, Switzerland
e-mail: alessio.figalli@math.ethz.ch

R. L. Frank
Department of Mathematics, LMU Munich, München, Germany

Munich Center for Quantum Science and Technology, München, Germany

Mathematics, Caltech, Pasadena, CA, USA
e-mail: r.frank@lmu.de

M. Loss
School of Mathematics, Georgia Institute of Technology, Atlanta, GA, USA
e-mail: loss@math.gatech.edu

V. S. Georgiev, T. Ozawa (eds.), *Various Methods for the Analysis of PDEs*,
ICIAM2023 Springer Series 5, https://doi.org/10.1007/978-981-96-9164-7_3

1 Introduction and Main Results

Let us assume that $d \geq 3$ without further notice. The classical *Sobolev inequality* on $\mathbb{R}^d$, can be written as follows:

$$\|\nabla f\|^2_{\mathrm{L}^2(\mathbb{R}^d)} \geq \mathsf{S}_d \, \|f\|^2_{\mathrm{L}^{2^*}(\mathbb{R}^d)} \quad \forall\, f \in \dot{\mathrm{H}}^1(\mathbb{R}^d)\,, \tag{S}$$

where $2^* = \frac{2d}{d-2}$ is the Sobolev exponent, $\mathsf{S}_d = \frac{1}{4}\, d\,(d-2)\,|\mathbb{S}^d|^{2/d}$ is the sharp Sobolev constant, and $|\mathbb{S}^d|$ denotes the volume of the unit sphere $\mathbb{S}^d \subset \mathbb{R}^{d+1}$. Here $\dot{\mathrm{H}}^1(\mathbb{R}^d)$ denotes the closure of $C^\infty_c(\mathbb{R}^d)$ with respect to the seminorm $|f|_{\dot{\mathrm{H}}^1(\mathbb{R}^d)} := \|\nabla f\|^2_{\mathrm{L}^2(\mathbb{R}^d)}$. As proved in [1, 42] (see also [40, 41]), equality in (S) holds if f is one of the *Aubin-Talenti* functions, that is, one of the functions belonging to the $(d+2)$-dimensional *Aubin-Talenti manifold*

$$\mathcal{M} := \left\{ g_{a,b,c} \,:\, (a,b,c) \in (0,+\infty) \times \mathbb{R}^d \times \mathbb{R} \right\}$$
$$\text{with} \quad g_{a,b,c}(x) := c\left(a + |x-b|^2\right)^{-\frac{d-2}{2}}.$$

In fact, there is equality in (S) if and only if f is in $\mathcal{M}$ according to [15, 32, 37].

Applying the inverse of the stereographic projection and integrating on the sphere $\mathbb{S}^d$ with respect to the uniform probability measure $d\mu_d$, one can rewrite the above inequality as

$$\|\nabla u\|^2_{\mathrm{L}^2(\mathbb{S}^d)} \geq \tfrac{1}{4}\, d\,(d-2)\left(\|u\|^2_{\mathrm{L}^{2^*}(\mathbb{S}^d)} - \|u\|^2_{\mathrm{L}^2(\mathbb{S}^d)}\right) \quad \forall\, u \in \mathrm{H}^1(\mathbb{S}^d) \tag{1}$$

and state, equivalently, that the only functions in $\mathrm{H}^1(\mathbb{S}^d)$ for which there is equality in (1) are the functions $\omega \mapsto G_{b,c}(\omega) := c\,(1 + b\cdot\omega)^{-(d-2)/2}$ where $b \in B_1 := \{b \in \mathbb{R}^{d+1} \,:\, |b| < 1\}$ and $c \in \mathbb{R}$ are constants. One can embed (1) in the following family of *Gagliardo-Nirenberg-Sobolev inequalities:*

$$\|\nabla u\|^2_{\mathrm{L}^2(\mathbb{S}^d)} \geq \frac{d}{p-2}\left(\|u\|^2_{\mathrm{L}^p(\mathbb{S}^d)} - \|u\|^2_{\mathrm{L}^2(\mathbb{S}^d)}\right) \quad \forall\, u \in \mathrm{H}^1(\mathbb{S}^d),\ \forall\, p \in (1,2)\cup(2,2^*]. \tag{GNS}$$

As proved by Bidaut-Véron and Véron [7] and Beckner [5] for $p > 2$ (see also [2–4] if $p \leq 2^\#$), the constant $d/(p-2)$ is the best possible constant in (GNS). Here $2^\# := (2\,d^2+1)/(d-1)^2 < 2^*$ denotes the Bakry-Emery exponent. If $p < 2^*$, the only optimizers in $\mathrm{H}^1(\mathbb{S}^d)$ for (GNS) are the constant functions. The *carré du champ* method used in [2–4] relies on the linear heat equation, which induces the limitation $p \leq 2^\#$. This limitation is not technical as shown in [27]. The whole range $p \in [1,2)\cup(2,2^*]$ was covered using a *carré du champ* method based on nonlinear diffusion equations by Demange and Dolbeault-Esteban-Loss in [22, 26].

The case $p = 2$ (that can be obtained from (GNS) by taking the limit $p \to 2$) is the *logarithmic Sobolev inequality*

$$\|\nabla u\|^2_{\mathrm{L}^2(\mathbb{S}^d)} \geq \frac{d}{2} \int_{\mathbb{S}^d} |u|^2 \ln\left(\frac{|u|^2}{\|u\|^2_{\mathrm{L}^2(\mathbb{S}^d)}}\right) d\mu_d \quad \forall\, u \in \mathrm{H}^1(\mathbb{S}^d) \setminus \{0\}\,, \tag{LSI}$$

where the only optimizers are the constant functions. Inequality (LSI) can also be proved by the *carré du champ* method based on the heat equation.

A remarkable feature of the *carré du champ* method is that additional terms appear in the computations. This has been used in [25] to establish, for all $p \in (2, 2^*)$, the *improved interpolation inequalities*

$$\begin{aligned} &\|\nabla u\|^2_{\mathrm{L}^2(\mathbb{S}^d)} - \frac{d}{p-2}\left(\|u\|^2_{\mathrm{L}^p(\mathbb{S}^d)} - \|u\|^2_{\mathrm{L}^2(\mathbb{S}^d)}\right) \\ &\geq \|u\|^2_{\mathrm{L}^2(\mathbb{S}^d)}\, \Psi_{d,p}\left(\frac{\|\nabla u\|^2_{\mathrm{L}^2(\mathbb{S}^d)}}{\|u\|^2_{\mathrm{L}^2(\mathbb{S}^d)}}\right) \quad \forall\, u \in \mathrm{H}^1(\mathbb{S}^d) \setminus \{0\} \end{aligned} \tag{2}$$

for some convex function $\Psi_{d,p}$ such that $\Psi_{d,p}(0) = \Psi'_{d,p}(0) = 0$ and $\Psi_{d,p}(s) > 0$ if $s > 0$. Inequality (2) is obviously stronger than (GNS) and can be used to prove that the equality case in (GNS) is realized only by constant functions. If we test this inequality with $u_\varepsilon(\omega) = 1 + \varepsilon\, \mathsf{b} \cdot \omega$ for a given $\mathsf{b} \in B_1$, an elementary computation shows that there is a cancellation of the $O(\varepsilon^2)$ terms as $\varepsilon \to 0$ and the first non-zero term is of the order of $O(\varepsilon^4)$.

If we denote by Π_1 the $\mathrm{L}^2(\mathbb{S}^d)$ projection on the space generated by the coordinate functions ω_i with $i = 1, 2, \ldots, d$, we learn from [10, Theorem 6] that there is an explicit constant $\kappa = \kappa(p, d) \in (0, 1)$, depending on d and $p \in [1, 2) \cup (2, 2^*)$ such that, for all $u \in \mathrm{H}^1(\mathbb{S}^d)$,

$$\begin{aligned} &\|\nabla u\|^2_{\mathrm{L}^2(\mathbb{S}^d)} - \frac{d}{p-2}\left(\|u\|^2_{\mathrm{L}^p(\mathbb{S}^d)} - \|u\|^2_{\mathrm{L}^2(\mathbb{S}^d)}\right) \\ &\geq \kappa\left(\frac{\|\nabla \Pi_1 u\|^4_{\mathrm{L}^2(\mathbb{S}^d)}}{\|\nabla u\|^2_{\mathrm{L}^2(\mathbb{S}^d)} + \|u\|^2_{\mathrm{L}^2(\mathbb{S}^d)}} + \|\nabla(\mathrm{Id} - \Pi_1)\, u\|^2_{\mathrm{L}^2(\mathbb{S}^d)}\right). \end{aligned} \tag{3}$$

The proof relies on (2) and on a decomposition in spherical harmonics directly inspired by [31]. Inequality (3) improves over (GNS), but this improvement degenerates as $p \to 2^*$. Indeed, any constant $\kappa(p, d)$ for which (3) holds necessarily satisfies $\lim_{p \to 2^*} \kappa(p, d) = 0$, for otherwise we had an inequality at $p = 2^*$ whose right side vanishes only for constants while, according to our discussion of (1), the left side vanishes for all function $G_{b,c}$.

Let us notice that Inequality (2) can take various forms. For instance, if $p \in (2, 2^\#)$ and $1/\theta = 1 + (\frac{d-1}{d+2})^2 \, (p-1) \, (2^\# - p)/(p-2)$, we read from [23, Ineq. (2.4)] that (2) takes the form

$$\|\nabla u\|^2_{\mathrm{L}^2(\mathbb{S}^d)} \geq \frac{d\,\theta}{p-2} \left(\|u\|^{2/\theta}_{\mathrm{L}^p(\mathbb{S}^d)} \, \|u\|^{2-2/\theta}_{\mathrm{L}^2(\mathbb{S}^d)} - \|u\|^2_{\mathrm{L}^2(\mathbb{S}^d)} \right) \quad \forall\, u \in \mathrm{H}^1(\mathbb{S}^d)\,,$$

which is a strict improvement compared to (GNS), as can be seen using Hölder inequalities. The case $p = 2$ is also covered, except that in the left-hand side of (2), the deficit of (GNS) has to be replaced by the deficit of (LSI).

Without entering into the details, let us quote some related results. By a direct variational approach, an improved inequality like (2) is proved in [31, Theorem 2], in the subcritical range, for some $\Psi_{d,p}(s) \sim s^2$. Using nonlinear flows and appropriate orthogonality constraints, improved inequalities with $\Psi_{d,p}(s) \sim s$ are known from [27]. Both results are unified in [10]. Let us finally mention that improved inequalities are proved in [8] with explicit constants, not on $\mathbb{S}^d$ but on $\mathbb{R}^d$, for the Gagliardo-Nirenberg-Sobolev inequalities using entropy methods and regularization effects for fast diffusion flows, but with some restrictions on the decay of the functions at infinity.

The Gaussian measure on $\mathbb{R}^N$ can be seen as an infinite dimensional limit $d \to +\infty$ of the uniform probability measure on the sphere of dimension d and radius $\sqrt{d}$ tested against functions depending only on a finite number N of coordinates: see for instance [11, 28, 39]. Since $\lim_{d\to+\infty} 2^* = 2$, it also turns out that (S) has to be replaced by a Gaussian interpolation inequality as follows. On $\mathbb{R}^N$, with $N \geq 1$, let us consider the *Gaussian measure* $d\gamma(x) := e^{-\pi\,|x|^2}\,dx$. With $\mathrm{L}^2(\gamma) := \mathrm{L}^2(\mathbb{R}^N, d\gamma)$, if $\mathrm{H}^1(\gamma)$ denotes the space of all $u \in \mathrm{L}^2(\gamma)$ with distributional gradient in $\mathrm{L}^2(\gamma)$, the *logarithmic Sobolev inequality* is:

$$\int_{\mathbb{R}^N} |\nabla u|^2\,d\gamma \geq \pi \int_{\mathbb{R}^N} |u|^2 \ln\left(\frac{|u|^2}{\|u\|^2_{\mathrm{L}^2(\gamma)}}\right) d\gamma \quad \forall\, u \in \mathrm{H}^1(\gamma) \setminus \{0\}\,. \tag{LSI}$$

According to the result of Carlen [16, Theorem 5], equality in (LSI) holds if and only if for some $a \in \mathbb{R}^N$ and $c \in \mathbb{R}$,

$$u(x) = c\,e^{a\cdot x}\,. \tag{4}$$

Improved forms of the inequality are also known for instance from [10, 24, 29, 30], under some restrictions. The article [33] makes a very interesting connection between deficit estimates for (LSI) (for certain restricted classes of functions) and the Mahler conjecture in convex geometry. A detailed list of earlier results and references can be found in [34].

Now let us turn our attention to *stability* issues for (S), (GNS) and (LSI).

- ***Stability for the Sobolev inequality***. In [9] Brezis and Lieb asked the following question:

  (Q) Do there exist constants $\kappa, \alpha > 0$ such that the Sobolev deficit $\boldsymbol{\delta}$ *controls some distance* $\mathbf{d}$ *from the Aubin-Talenti manifold* $\mathcal{M}$ *according to*

$$\delta(f) := \|\nabla f\|^2_{\mathrm{L}^2(\mathbb{R}^d)} - \mathsf{S}_d\, \|f\|^2_{\mathrm{L}^{2^*}(\mathbb{R}^d)} \geq \kappa\, \mathbf{d}(f, \mathcal{M})^\alpha \quad ?$$

  The 'best possible answer' to this question would involve finding the strongest possible topology to define the distance $\mathbf{d}$ and the best possible constant κ and exponent α. The first answer to Brezis and Lieb's question was given by Bianchi and Egnell in [6]: there is a constant $C_{d,\mathrm{BE}} > 0$ such that

$$\mathbf{d}(f, \mathcal{M})^2 := \inf_{g \in \mathcal{M}} \|\nabla f - \nabla g\|^2_{\mathrm{L}^2(\mathbb{R}^d)} \,. \tag{5}$$

  Similar results for other inequalities have been proved using the strategy of Bianchi and Egnell: see, for example, [19]. The main drawback of this strategy is that no explicit estimate of $C_{d,\mathrm{BE}}$ is known nor its dependence on the dimension d. Recently, in [28], we proved the following result.

Theorem 1 ([28, Theorem 1.1]) *There is an explicit constant $\beta > 0$ such that for any $d \geq 3$,*

$$\|\nabla f\|^2_{\mathrm{L}^2(\mathbb{R}^d)} - \mathsf{S}_d\, \|f\|^2_{\mathrm{L}^{2^*}(\mathbb{R}^d)} \geq \frac{\beta}{d} \mathbf{d}(f, \mathcal{M})^2 \quad \forall\, f \in \dot{\mathrm{H}}^1(\mathbb{R}^d)\,. \tag{6}$$

  This result is dimensionally sharp. Indeed, Theorem 1 can be rewritten as $C_{d,\mathrm{BE}} \geq \beta/d$. On the other hand, it was proved implicitly in [6] (see also [19]) that $C_{d,\mathrm{BE}} \leq 4/(4+d)$. This inequality is in fact strict: $C_{d,\mathrm{BE}} < 4/(4+d)$, according to [35], and we learn from [36] that equality in (5), written with the optimal value of $C_{d,\mathrm{BE}}$, is achieved. Hence, Theorem 1 captures the dimensional behavior of $C_{d,\mathrm{BE}}$. For completeness, let us mention the extension of Theorem 1 in [20] to fractional Sobolev inequalities.

  Using the inverse stereographic projection, the stability result of Theorem 1 can be rewritten on $\mathbb{S}^d$ for any $u \in \mathrm{H}^1(\mathbb{S}^d)$ as

$$\begin{aligned}
&\|\nabla u\|^2_{\mathrm{L}^2(\mathbb{S}^d)} - \tfrac{1}{4}\, d\,(d-2) \left(\|u\|^2_{\mathrm{L}^{2^*}(\mathbb{S}^d)} - \|u\|^2_{\mathrm{L}^2(\mathbb{S}^d)}\right) \\
&\geq \frac{\beta}{d} \inf_{(b,c) \in B_1 \times \mathbb{R}} \left(\left\|\nabla u - \nabla G_{b,c}\right\|^2_{\mathrm{L}^2(\mathbb{S}^d)} + \tfrac{1}{4}\, d\,(d-2)\, \left\|u - G_{b,c}\right\|^2_{\mathrm{L}^2(\mathbb{S}^d)}\right) .
\end{aligned}$$

- ***Stability for the Gaussian logarithmic Sobolev inequality***. The interpretation of the Gaussian measure as the limit of uniform probability measures on d-dimensional spheres as $d \to +\infty$ and the explicit dimensional dependence of the stability constant of Theorem 1 provides us with a stability result for (LSI).

Theorem 2 ([28, Corollary 1.2]) *With $\beta > 0$ as in Theorem 1, for all $N \in \mathbb{N}$, we have*

$$\int_{\mathbb{R}^N} |\nabla u|^2 \, d\gamma - \pi \int_{\mathbb{R}^N} |u|^2 \ln\left(\frac{|u|^2}{\|u\|^2_{\mathrm{L}^2(\gamma)}}\right) d\gamma$$
$$\geq \frac{\beta\,\pi}{2} \inf_{b\in\mathbb{R}^N,\, c\in\mathbb{R}} \int_{\mathbb{R}^N} \left|u - c\, e^{b\cdot x}\right|^2 d\gamma \quad \forall\, u \in \mathrm{H}^1(\gamma)\,. \tag{7}$$

In [28] this result was obtained as a corollary of Theorem 1. In this paper we give a new, direct proof which highlights the strategy of [28] in a slightly simpler setting.

In the next sections, we briefly describe the strategies of [25] and [28] to prove the improvements in the case of subcritical inequalities (Sect. 2) and the stability results of Theorem 1 (Sect. 3) and Theorem 2 (Sect. 4). The new proof of Theorem 2 is given in Sect. 5.

2 Improved Interpolation Inequalities

In this section we describe the method used in [25] to prove (2) and similar improvements of the (GNS) inequalities in the subcritical case $p \in (2, 2^*)$. Inequalities (GNS) and their limiting case corresponding to $p = 2$ can be written as:

$\mathsf{i} \geq d\,\mathsf{e}$ where

$$\mathsf{i} := \|\nabla u\|^2_{\mathrm{L}^2(\mathbb{S}^d)}\,, \qquad \mathsf{e} := \begin{cases} \frac{1}{p-2}\left(\|u\|^2_{\mathrm{L}^p(\mathbb{S}^d)} - \|u\|^2_{\mathrm{L}^2(\mathbb{S}^d)}\right) & \text{if} \quad p \neq 2\,, \\ \frac{1}{2}\int_{\mathbb{S}^d} |u|^2 \ln\left(\frac{|u|^2}{\|u\|^2_{\mathrm{L}^2(\mathbb{S}^d)}}\right) d\mu_d & \text{if} \quad p = 2\,, \end{cases}$$

for any $u \in \mathrm{H}^1(\mathbb{S}^d)$. By homogeneity we can assume that $\|u\|_{\mathrm{L}^2(\mathbb{S}^d)} = 1$. Let $I_p := [0, +\infty)$ if $p \geq 2$ and $I_p := [0, 1/(2-p))$ if $p \in (1, 2)$.

Theorem 3 ([25, Theorem 1.1]) *Let $d \geq 3$ and $p \in (2, 2^*)$. With the above notation and conventions, there is an explicit function φ, such that $\varphi(0) = 0$, $\varphi'(0) = 1$, and $\varphi'' > 0$ on I_p, for which*

$$\mathsf{i} \geq d\,\varphi(\mathsf{e})\,. \tag{8}$$

Similar results can be proved in dimension $d = 1$ and $d = 2$: see [10, 22, 23, 25]. Since

$$\mathsf{i} - d\,\mathsf{e} \geq d\left(\varphi(\mathsf{e}) - \mathsf{e}\right) \geq 0\,,$$

it is clear that the equality case in (2) holds if and only if $\mathsf{e} = 0$, that is, if u is a constant. Moreover $\varphi(\mathsf{e}) - \mathsf{e}$ measures a distance to the constants, for instance in $\mathrm{L}^1(\mathbb{S}^d)$ norm using a generalized Ciszár-Kullback-Pinsker inequality, and $\varphi(s)-s \sim \varphi''(0)\,s^2/2$ as $s \to 0$. To prove (2) in the stronger homogeneous Sobolev norm $\dot{\mathrm{H}}^1(\mathbb{R}^d)$, it is enough to define the convex function

$$\Psi_{d,p}(s) := s - d\,\varphi^{-1}\left(\frac{s}{d}\right).$$

It is elementary to verify that $\Psi_{d,p}(0) = \Psi'_{d,p}(0) = 0$ and that $\Psi_{d,p}(s) > 0$ if $s \neq 0$.

Let us give a sketch of the proof of Theorem 3 based on the method of [25]. As a first step, using Schwarz foliated symmetrization (see for instance [25, Section 2] for references) and cylindrical coordinates on $\mathbb{S}^d \subset \mathbb{R}^{d+1}$, we can reduce the problem to functions depending only on one coordinate $z \in [-1, 1]$ corresponding to the South Pole–North Pole axis. In other words, we consider a function $u(\omega) = f(z)$, where $\omega = (\omega_1, \omega_2, \ldots, \omega_{d+1})$ and $z = \omega_{d+1}$, and find that

$$\|\nabla u\|^2_{\mathrm{L}^2(\mathbb{S}^d)} = \int_{-1}^{1} |f'(z)|^2 \left(1-z^2\right) d\sigma_d(z) \quad \text{and} \quad \|u\|^q_{\mathrm{L}^q(\mathbb{S}^d)} = \int_{-1}^{1} |f(z)|^q\, d\sigma_d(z)\,,$$

where $'$ denotes the z-derivative and where

$$d\sigma_d(z) := \frac{\Gamma(\frac{d+1}{2})}{\sqrt{\pi}\,\Gamma(\frac{d}{2})}\left(1 - z^2\right)^{\frac{d}{2}-1} dz\,.$$

In particular, the Laplace-Beltrami operator on $\mathbb{S}^d$ is reduced to the *ultraspherical* operator: $\Delta u = \mathcal{L} f := \left(1 - z^2\right) f'' - d\,z\,f'$ on $(-1, 1)$.

The key idea is to prove that $\mathsf{i} - d\,\varphi(\mathsf{e})$ is monotone nonincreasing under the action of

$$\frac{\partial \rho}{\partial t} = \mathcal{L}\,\rho^m\,, \tag{9}$$

where $m = 1$ corresponds to the heat flow, $m > 1$ to the porous medium flow and $m < 1$ to the fast diffusion flow. Here we choose $\rho = |f|^p$ so that $\|u\|_{\mathrm{L}^p(\mathbb{R}^d)}$ is conserved under the action of (9). It is convenient to introduce the exponent β such that

$$m = 1 + \tfrac{2}{p}\left(\tfrac{1}{\beta} - 1\right)$$

and consider the function $f = w^\beta$, such that $w^{\beta\,p} = \rho$, which solves

$$\frac{\partial w}{\partial t} = w^{2-2\beta}\left(\mathcal{L}w + \kappa\,\frac{|w'|^2}{w}\right) \quad \text{with} \quad \kappa := \beta\,(p-2)+1\,. \tag{10}$$

The *carré du champ* method shows that $\frac{d}{dt}\,(\mathsf{i} - d\,\mathsf{e}) \le 0$ if the function

$$\gamma(\beta) := -\left(\tfrac{d-1}{d+2}\,(\kappa+\beta-1)\right)^2 + \kappa\,(\beta-1) + \tfrac{d}{d+2}\,(\kappa+\beta-1)$$

takes nonnegative values, which amounts to $m_-(d,p) \le m \le m_+(d,p)$ where

$$m_\pm(d,p) := \frac{1}{(d+2)\,p}\left(d\,p + 2 \pm \sqrt{d\,(d-2)\,(p-1)\,(2^*-p)}\right).$$

Using $\lim_{t\to+\infty}(\mathsf{i}(t) - d\,\mathsf{e}(t)) = 0$, we conclude that $\mathsf{i}(t) - d\,\mathsf{e}(t) \ge 0$ for any $t \ge 0$ and, as a special case, for $t = 0$: this proves (GNS) for an arbitrary initial datum. If $m_-(d,p) < m < m_+(d,p)$, we find that $\gamma(\beta) > 0$, which leaves some space for an improvement. A more detailed and quite lengthy computation along the flow (10) yields

$$\frac{d}{dt}\big(\mathsf{i} - d\,\varphi(\mathsf{e})\big) \le \frac{\gamma}{\beta^2}\,\frac{\mathsf{i} - d\,\varphi(\mathsf{e})}{\big(1-(p-2)\,\mathsf{e}\big)^\delta}\,\frac{d\mathsf{e}}{dt},$$

where $\delta = 1$ if $1 \le p \le 2$ and $\delta := \frac{2-(4-p)\,\beta}{2\,\beta\,(p-2)}$ if $p > 2$, if φ solves

$$\frac{d\varphi}{ds} = 1 + \frac{\gamma}{\beta^2}\,\frac{\varphi(s)}{(1\;-\;(p-2)\,s)^\delta}\,, \qquad \varphi(0) = 0\,.$$

This proves (8) for any $t \ge 0$ as a consequence of the monotonicity of $\mathsf{i} - d\,\varphi(\mathsf{e})$ and the fact that $\lim_{t\to+\infty}\big(\mathsf{i}(t) - d\,\varphi(\mathsf{e}(t))\big) = 0$. See [10, Appendix B.4] for detailed justifications. □

As a consequence of the *carré du champ* method, optimal constants in Gagliardo-Nirenberg-Sobolev inequalities determine optimal rates of decay in some associated fast diffusion equations. A stability result for a functional inequality can usually be rephrased as an improved inequality, under appropriate constraints. Measured by the dissipation of some entropy functional, this means a faster convergence rate towards an asymptotic profile for solutions of the associated evolution equation: see [17] for an application of stability for Gagliardo-Nirenberg-Sobolev inequalities to the proof of a quantitative rate of convergence for solutions of the Keller-Segel system, and [10] for explicit, improved, convergence rates based on (GNS) in the subcritical regime, in the case of the sphere. On the Euclidean space, improved convergence rates based on stability for Gagliardo-Nirenberg-Sobolev inequalities (subcritical and critical cases) and entropy methods are studied in [8]. Also see [12, 30] for the limiting case of the logarithmic Sobolev inequality and corresponding improved decay rates for linear heat flows. Notice that stability estimates can also be applied in the perspective of estimating a posteriori errors both from theoretical (test functions) and numerical (approximate numerical solutions) points of view.

3 Stability of the Sobolev Inequality: Proof of Theorem 1

In this section we explain the general ideas of the proof of Theorem 1 in [28].

3.1 On the Stability Proof by Bianchi and Egnell

The strategy of Bianchi-Egnell to prove (5) is based on two main steps:

(1) A local stability estimate in a neighborhood of $\mathcal{M}$, obtained by a local spectral analysis.
(2) A reduction of the global estimate to the local estimate by the concentration-compactness method based on Lions' analysis (see [38]).

Theorem 1 is a significant improvement of Bianchi-Egnell's result, as it contains a dimensionally sharp lower estimate for the best stability constant. Our strategy is to make the *second step constructive* and the *first one explicit,* with much more detailed estimates.

3.2 Strategy of the Proof of Theorem 1

The proof is divided into several steps:

(1) Local analysis: prove the inequality for *nonnegative* functions close to $\mathcal{M}$ with an explicit remainder term. The analysis is quite involved: it relies on "cuttings" at various heights, the use of uniform bounds on spherical harmonics and some delicate concavity properties.
(2) Local to global extension: prove the inequality for *nonnegative* functions far from $\mathcal{M}$ using the *competing symmetries* of [18] and a continuous Steiner symmetrization.
(3) Deduce the inequality for *sign-changing* functions from the inequality for *nonnegative* functions by a concavity argument.

- In the first step, in order to obtain uniform estimates as $d \to +\infty$, we need to expand

$$(1+r)^{2^*} - 1 - 2^* r,$$

with an accurate remainder term, for all $r \geq -1$. To do that, we "cut r into pieces" by defining

$$r_1 := \min\{r, \gamma\}, \quad r_2 := \min\left\{(r-\gamma)_+, M-\gamma\right\} \quad \text{and} \quad r_3 := (r-M)_+ ,$$

where γ and M are suitable parameters satisfying $0 < \gamma < M$. Furthermore, define

$$\theta := 2^* - 2 = \tfrac{4}{d-2} \, .$$

Notice that $\theta \in (0, 1]$ if $d \geq 6$ and $\lim_{d\to+\infty} \theta(d) = 0$.

Lemma 4 ([28, Proposition 2.9]) *Given $d \geq 6$, $r \in [-1, \infty)$, and $\overline{M} \in [\sqrt{e}, +\infty)$, we have*

$$(1+r)^{2^*} - 1 - 2^* r \leq \tfrac{1}{2}\, 2^*\, (2^*-1)\, (r_1+r_2)^2 + 2\, (r_1+r_2)\, r_3 + \left(1 + C_M\, \theta\, \overline{M}^{-1} \ln \overline{M}\right) r_3^{2^*}$$
$$+ \left(\tfrac{3}{2}\, \gamma\, \theta\, r_1^2 + C_{M,\overline{M}}\, \theta\, r_2^2\right) \mathbb{1}_{\{r \leq M\}} + C_{M,\overline{M}}\, \theta\, M^2\, \mathbb{1}_{\{r > M\}} ,$$

where all the constants in the above inequality are explicit.

One can then prove that there exist computable constants ϵ_1, ϵ_2, k_0, and $\epsilon_0 \in (0, 1/\theta)$, such that

$$\|\nabla r\|^2_{\mathrm{L}^2(\mathbb{S}^d)} + \mathrm{A}\, \|r\|^2_{\mathrm{L}^2(\mathbb{S}^d)} - \mathrm{A}\, \|1+r\|^2_{\mathrm{L}^{2^*}(\mathbb{S}^d)}$$
$$\geq \frac{4\, \epsilon_0}{d-2} \left(\|\nabla r\|^2_{\mathrm{L}^2(\mathbb{S}^d)} + \mathrm{A}\, \|r\|^2_{\mathrm{L}^2(\mathbb{S}^d)}\right) + \sum_{k=1}^{3} I_k ,$$

with $\mathrm{A} := \frac{1}{4}\, d\, (d-2)$ and

$$I_1 := (1-\theta\, \epsilon_0) \textstyle\int_{\mathbb{S}^d} \left(|\nabla r_1|^2 + \mathrm{A}\, r_1^2\right) d\mu_d$$
$$- \mathrm{A}\, (2^* - 1 + \epsilon_1\, \theta) \int_{\mathbb{S}^d} r_1^2\, d\mu_d + \mathrm{A}\, k_0\, \theta \int_{\mathbb{S}^d} (r_2^2 + r_3^2)\, d\mu_d ,$$

$$I_2 := (1-\theta\, \epsilon_0) \textstyle\int_{\mathbb{S}^d} \left(|\nabla r_2|^2 + \mathrm{A}\, r_2^2\right) d\mu_d - \mathrm{A}\left(2^* - 1 + (k_0 + C_{\epsilon_1,\epsilon_2})\, \theta\right) \textstyle\int_{\mathbb{S}^d} r_2^2\, d\mu_d ,$$

$$I_3 := (1-\theta\, \epsilon_0) \textstyle\int_{\mathbb{S}^d} \left(|\nabla r_3|^2 + \mathrm{A}\, r_3^2\right) d\mu_d$$
$$- \tfrac{2}{2^*}\, \mathrm{A}\, (1 + \epsilon_2\, \theta) \int_{\mathbb{S}^d} r_3^{2^*}\, d\mu_d - \mathrm{A}\, k_0\, \theta \int_{\mathbb{S}^d} r_3^2\, d\mu_d .$$

Next, one can use spectral gap estimates to prove $I_1 \geq 0$ and the Sobolev inequality to prove $I_3 \geq 0$, noting that the extra coefficient $2/2^* < 1$ gives enough room to accommodate all error terms. Finally, using that $\mu\big(\{r_2 > 0\}\big)$ is small, an improved spectral gap inequality allows us to show $I_2 \geq 0$.

If $d = 3$, 4, or 5, we can rely on a simpler Taylor expansion that can be found in [28, Proposition 2.7]. As a consequence, the following result has been proved.

Theorem 5 ([28, Theorem 2.1]) *There are explicit constants $\epsilon_0 \in (0, 1/3)$ and $\delta \in (0, 1/2)$ such that for all $d \geq 3$ and for all nonnegative $u = 1 + r \in \mathrm{H}^1(\mathbb{S}^d)$ with*

$$\|r\|^2_{\mathrm{L}^{2^*}(\mathbb{S}^d)} \leq \frac{\delta}{1-\delta}\,, \quad \int_{\mathbb{S}^d} r\,d\mu_d = 0 \quad and \quad \int_{\mathbb{S}^d} \omega\, r\,d\mu_d = 0\,,$$

one has

$$\|\nabla u\|^2_{\mathrm{L}^2(\mathbb{S}^d)} + \mathrm{A}\;\|u\|^2_{\mathrm{L}^2(\mathbb{S}^d)} - \mathrm{A}\;\|u\|^2_{\mathrm{L}^{2^*}(\mathbb{S}^d)} \geq \frac{4\,\epsilon_0}{d-2}\left(\|\nabla r\|^2_{\mathrm{L}^2(\mathbb{S}^d)} + \mathrm{A}\;\|r\|^2_{\mathrm{L}^2(\mathbb{S}^d)}\right).$$

Let us define the stability quotient

$$\mathcal{E}(f) := \frac{\|\nabla f\|^2_{\mathrm{L}^2(\mathbb{R}^d)} - \mathsf{S}_d\,\|f\|^2_{\mathrm{L}^{2^*}(\mathbb{R}^d)}}{\mathbf{d}(f,\mathcal{M})^2}$$

and consider the infimum

$$\mathscr{I}(\delta) := \inf\Big\{\mathcal{E}(f)\;:\;0 \leq f \in \dot{\mathrm{H}}^1(\mathbb{R}^d)\setminus\mathcal{M}\,,\quad \mathbf{d}(f,\mathcal{M})^2 \leq \delta\,\|\nabla f\|^2_{\mathrm{L}^2(\mathbb{R}^d)}\Big\}\,.$$

For a given f, up to a conformal transformation, we can assume that $\mathbf{d}(f,\mathcal{M})^2$ is realized by the Aubin-Talenti function $g = g_\star$ with

$$g_\star(x) := |\mathbb{S}^d|^{-\frac{d-2}{2d}}\left(\frac{2}{1+|x|^2}\right)^{\frac{d-2}{2}} \quad \forall\,x \in \mathbb{R}^d\,, \tag{11}$$

use the inverse stereographic projection to transform f and g respectively into $u = 1 + r$ and 1, and notice that

$$\mathrm{A}\;\|r\|^2_{\mathrm{L}^{2^*}(\mathbb{S}^d)} \leq \|\nabla r\|^2_{\mathrm{L}^2(\mathbb{S}^d)} + \mathrm{A}\;\|r\|^2_{\mathrm{L}^2(\mathbb{S}^d)} = \mathbf{d}(f,\mathcal{M})^2\,,$$

$$\|\nabla f\|^2_{\mathrm{L}^2(\mathbb{R}^d)} = \|\nabla u\|^2_{\mathrm{L}^2(\mathbb{S}^d)} + \mathrm{A}\;\|u\|^2_{\mathrm{L}^2(\mathbb{S}^d)} = \|\nabla r\|^2_{\mathrm{L}^2(\mathbb{S}^d)} + \mathrm{A}\;\|r\|^2_{\mathrm{L}^2(\mathbb{S}^d)} + \mathrm{A}\,,$$

where the first line follows from (1). We deduce from the condition $\mathbf{d}(f,\mathcal{M})^2 \le \delta\,\|\nabla f\|^2_{\mathrm{L}^2(\mathbb{R}^d)}$ that

$$\mathrm{A}\,\|r\|^2_{\mathrm{L}^{2^*}(\mathbb{S}^d)} \le \|\nabla r\|^2_{\mathrm{L}^2(\mathbb{S}^d)} + \mathrm{A}\,\|r\|^2_{\mathrm{L}^2(\mathbb{S}^d)} \le \frac{\delta\,\mathrm{A}}{1-\delta}$$

and apply Theorem 5 to obtain

$$\mathcal{E}(f) = \frac{\|\nabla u\|^2_{\mathrm{L}^2(\mathbb{S}^d)} + \mathrm{A}\,\|u\|^2_{\mathrm{L}^2(\mathbb{S}^d)} - \mathrm{A}\,\|u\|^2_{\mathrm{L}^{2^*}(\mathbb{S}^d)}}{\|\nabla r\|^2_{\mathrm{L}^2(\mathbb{S}^d)} + \mathrm{A}\,\|r\|^2_{\mathrm{L}^2(\mathbb{S}^d)}} \ge \frac{4\,\epsilon_0}{d-2}\,.$$

This completes the *local* analysis where, by homogeneity, the scale is fixed in terms of $\|\nabla f\|^2_{\mathrm{L}^2(\mathbb{R}^d)}$. With the notation of Theorem 5,

$$\mathscr{I}(\delta) \ge \frac{4\,\epsilon_0}{d-2}\,.$$

- In Step 2, we deal with nonnegative functions f that are not close to the manifold $\mathcal{M}$, i.e., such that

$$\mathbf{d}(f,\mathcal{M})^2 > \delta\,\|\nabla f\|^2_{\mathrm{L}^2(\mathbb{R}^d)}\,. \tag{12}$$

  The first ingredient is the method of *competing symmetries* [18] of Carlen and Loss. Consider any nonnegative function $f \in \dot{\mathrm{H}}^1(\mathbb{R}^d)$ and let

$$(Uf)(x) := \left(\frac{2}{|x-e_d|^2}\right)^{\frac{d-2}{2}} \times f\left(\frac{x_1}{|x-e_d|^2},\dots,\frac{x_{d-1}}{|x-e_d|^2},\frac{|x|^2-1}{|x-e_d|^2}\right)$$
$$\text{where}\quad e_d = (0,\dots,0,1) \in \mathbb{R}^d\,,$$

  and notice that $\mathcal{E}(Uf) = \mathcal{E}(f)$. We also consider the symmetric decreasing rearrangement $\mathcal{R}f = f^*$, with the properties that f and f^* are equimeasurable, and that $\|\nabla f^*\|_{\mathrm{L}^2(\mathbb{R}^d)} \le \|\nabla f\|_{\mathrm{L}^2(\mathbb{R}^d)}$. The following result is taken from the proof of [18, Theorem 3.3].

Theorem 6 ([18]) *Let $f \in \mathrm{L}^{2^*}(\mathbb{R}^d)$ be a nonnegative function with $\|f\|_{\mathrm{L}^{2^*}(\mathbb{R}^d)}=1$. The sequence $f_n = (\mathcal{R}U)^n f$ is such that* $\lim_{n\to+\infty}\|f_n - g_*\|_{\mathrm{L}^{2^*}(\mathbb{R}^d)} = 0$. *If $f \in \dot{\mathrm{H}}^1(\mathbb{R}^d)$, then $(\|\nabla f_n\|_{\mathrm{L}^2(\mathbb{R}^d)})_{n\in\mathbb{N}}$ is a nonincreasing sequence.*

Whether f_n satisfies (12) for all $n \in \mathbb{N}$ or not, we face an alternative.

Lemma 7 ([28, Lemma 3.5]) *Let f be as in Theorem 6 and such that* (12) *holds and let $f_n = (\mathcal{R}U)^n f$. Then either $\mathbf{d}(f_n, \mathcal{M})^2 \geq \delta \, \|\nabla f_n\|^2_{\mathrm{L}^2(\mathbb{R}^d)}$ for all $n \in \mathbb{N}$, or there exists $n_0 \in \mathbb{N}$ such that*

$$\mathbf{d}(f_{n_0}, \mathcal{M})^2 > \delta \, \|\nabla f_{n_0}\|^2_{\mathrm{L}^2(\mathbb{R}^d)} \quad \textit{and}$$

$$\mathbf{d}(f_{n_0+1}, \mathcal{M})^2 < \delta \, \|\nabla f_{n_0+1}\|^2_{\mathrm{L}^2(\mathbb{R}^d)} \,.$$

In the first case, that is if f_n satisfies (12) for all $n \in \mathbb{N}$, we have

$$\lim_{n\to+\infty} \|\nabla f_n\|_2^2 \leq \frac{1}{\delta} \lim_{n\to+\infty} \mathbf{d}(f_n, \mathcal{M})^2 = \frac{1}{\delta} \left(\lim_{n\to+\infty} \|\nabla f_n\|_2^2 - S_d \, \|f\|_{2^*}^2 \right),$$

where the last equality arises as a consequence of the properties of $(f_n)_{n\in\mathbb{N}}$ (see [28, Lemma 3.4]). Combined with the simple estimate

$$\mathcal{E}(f) = \frac{\|\nabla f\|_2^2 - S_d \, \|f\|_{2^*}^2}{\mathbf{d}(f, \mathcal{M})^2} \geq \frac{\|\nabla f\|_2^2 - S_d \, \|f\|_{2^*}^2}{\|\nabla f\|_2^2} \geq \frac{\|\nabla f_n\|_2^2 - S_d \, \|f\|_{2^*}^2}{\|\nabla f_n\|_2^2},$$

we can take the limit as $n \to +\infty$ and obtain $\mathcal{E}(f) \geq \delta$. In the second case, we adapt a strategy that has some similarities to one used by Christ in [21] and is based on a continuous rearrangement *flow* $(\mathsf{f}_\tau)_{n_0 \leq \tau < n_0+1}$ with $\mathsf{f}_{n_0} = U f_{n_0}$ such that

$$\|\mathsf{f}_\tau\|_{\mathrm{L}^{2^*}(\mathbb{R}^d)} = \|\mathsf{f}_{n_0}\|_{\mathrm{L}^{2^*}(\mathbb{R}^d)}\,, \quad \tau \mapsto \|\nabla \mathsf{f}_\tau\|_{\mathrm{L}^2(\mathbb{R}^d)} \text{ is nonincreasing, and}$$

$$\lim_{\tau\to n_0+1} \mathsf{f}_\tau = f_{n_0+1} \,.$$

Choosing the smallest $\tau_0 \in (n_0, n_0+1)$ such that $\mathbf{d}(f_{\tau_0}, \mathcal{M})^2 = \delta \, \|\nabla \mathsf{f}_{\tau_0}\|^2_{\mathrm{L}^2(\mathbb{R}^d)}$ and using $\|f\|_{\mathrm{L}^{2^*}(\mathbb{R}^d)} = \|\mathsf{f}_{n_0}\|_{\mathrm{L}^{2^*}(\mathbb{R}^d)} = \|\mathsf{f}_{\tau_0}\|_{\mathrm{L}^{2^*}(\mathbb{R}^d)}$, this gives:

$$1 - S_d \, \frac{\|f\|^2_{\mathrm{L}^{2^*}(\mathbb{R}^d)}}{\|\nabla f\|^2_{\mathrm{L}^2(\mathbb{R}^d)}} \geq 1 - S_d \, \frac{\|\mathsf{f}_{\tau_0}\|^2_{\mathrm{L}^{2^*}(\mathbb{R}^d)}}{\|\nabla \mathsf{f}_{\tau_0}\|^2_{\mathrm{L}^2(\mathbb{R}^d)}} = \delta \, \mathcal{E}(\mathsf{f}_{\tau_0}) \geq \delta \, \mathscr{I}(\delta) \,.$$

As a consequence, we obtain $C^{\mathrm{pos}}_{d,\mathrm{BE}} := \inf_{f\geq 0} \mathcal{E}(f) \geq \min\{\delta, \delta \, \mathscr{I}(\delta)\} = \delta \, \mathscr{I}(\delta)$ because $\mathscr{I}(\delta) < 1$. To build the flow, we refer to [13, 14] and to [28] for further references. The existence of τ_0 requires a discussion that can be found in [28, Section 3.1.2].

- The third step is to remove the positivity assumption of Theorem 1 as in [28, Section 3.2]. Take $f = f_+ - f_-$ with $\|f\|_{\mathrm{L}^{2^*}(\mathbb{R}^d)} = 1$ and define $m := \|f_-\|^{2^*}_{\mathrm{L}^{2^*}(\mathbb{R}^d)}$. Without loss of generality one may assume that $1 - m = \|f_+\|^{2^*}_{\mathrm{L}^{2^*}(\mathbb{R}^d)} > 1/2$. The positive concave function

$$h_d(m) := m^{\frac{d-2}{d}} + (1-m)^{\frac{d-2}{d}} - 1$$

satisfies

$$2\,h_d(1/2)\,m \le h_d(m)\,, \quad h_d(1/2) = 2^{2/d} - 1\,.$$

With $\boldsymbol{\delta}(f) = \|\nabla f\|^2_{\mathrm{L}^2(\mathbb{R}^d)} - \mathsf{S}_d\,\|f\|^2_{\mathrm{L}^{2^*}(\mathbb{R}^d)}$, one finds $g_+ \in \mathcal{M}$ such that

$$\boldsymbol{\delta}(f) \ge C^{\mathrm{pos}}_{d,\mathrm{BE}}\,\|\nabla f_+ - \nabla g_+\|^2_{\mathrm{L}^2(\mathbb{R}^d)} + \frac{2\,h_d(1/2)}{h_d(1/2)+1}\,\|\nabla f_-\|^2_{\mathrm{L}^2(\mathbb{R}^d)}\,,$$

and therefore

$$C_{d,\mathrm{BE}} \ge \tfrac{1}{2}\,\min\left\{\delta\,\mathscr{I}(\delta),\,\frac{2\,h_d(1/2)}{h_d(1/2)+1}\right\}\,.$$

- Combining all estimates of the three previous steps completes the proof of Theorem 1. □

4 Stability of the Gaussian Logarithmic Sobolev Inequality: Proof of Theorem 2

In this section we describe the main steps in the proof of the stability estimate for the Gaussian logarithmic Sobolev inequality (LSI) by considering the large dimensional limit of (6), as it appears in [28, Section 4]. With $u = f/g_\star$, where $g_\star$ is given by (11), Inequality (6) can be rewritten as

$$\begin{aligned}&\int_{\mathbb{R}^d}|\nabla u|^2\,g_\star^2\,dx + d\,(d-2)\int_{\mathbb{R}^d}|u|^2\,g_\star^{2^*}\,dx - d\,(d-2)\,\|g_\star\|^{2^*-2}_{\mathrm{L}^{2^*}(\mathbb{R}^d)}\\ &\quad\times\left(\int_{\mathbb{R}^d}|u|^{2^*}\,g_\star^{2^*}\,dx\right)^{2/2^*}\\ &\ge\frac{\beta}{d}\left(\int_{\mathbb{R}^d}|\nabla u|^2\,g_\star^2\,dx + d\,(d-2)\int_{\mathbb{R}^d}|u-g_d/g_\star|^2\,g_\star^{2^*}\,dx\right),\end{aligned}$$

where $g_d \in \mathcal{M}$ realizes $\mathbf{d}(f, \mathcal{M})^2$. For some parameters a_d, b_d and c_d, we can write that $g_d(x) = c_d\,(a_d + |x - b_d|^2)^{1-d/2}$. We rescale the function u according to

$$u(x) = v\,(r_d\,x) \quad \forall\,x \in \mathbb{R}^d\,, \quad r_d = \sqrt{\tfrac{d}{2\pi}}$$

and consider the function w_v^d such that $w_v^d\,(r_d\,x) = g_d(x)/g_\star(x)$. Hence

$$\begin{aligned}
&\int_{\mathbb{R}^d} |\nabla v|^2 \left(1 + \tfrac{1}{r_d^2}\,|x|^2\right)^2 d\boldsymbol{\mu}_d \\
&\geq 2\,\pi\,(d-2)\left(\left(\int_{\mathbb{R}^d} |v|^{2^*}\,d\boldsymbol{\mu}_d\right)^{2/2^*} - \int_{\mathbb{R}^d} |v|^2\,d\boldsymbol{\mu}_d\right) \\
&\quad + \frac{\beta}{d}\left(\int_{\mathbb{R}^d} |\nabla v|^2\,d\boldsymbol{\mu}_d + 2\,\pi(d-2)\int_{\mathbb{R}^d} |v - w_v^d|^2\,d\boldsymbol{\mu}_d\right),
\end{aligned}$$

where $d\boldsymbol{\mu}_d = Z_d^{-1}\,g_\star^{2^*}\,dx$ is the probability measure given by

$$d\boldsymbol{\mu}_d(x) := \frac{1}{Z_d}\left(1 + \tfrac{1}{r_d^2}\,|x|^2\right)^{-d} dx \quad \text{with} \quad Z_d = \frac{2^{1-d}\,\sqrt{\pi}}{\Gamma\left(\frac{d+1}{2}\right)}\left(\frac{d}{2}\right)^{\frac{d}{2}}.$$

Our goal is to take the limit $d \to +\infty$ when one considers functions $v(x)$ depending only on $y \in \mathbb{R}^N$, with $x = (y, z) \in \mathbb{R}^N \times \mathbb{R}^{d-N} \approx \mathbb{R}^d$, for some fixed integer N. With $|x|^2 = |y|^2 + |z|^2$, notice that

$$1 + \tfrac{1}{r_d^2}\,|x|^2 = 1 + \tfrac{1}{r_d^2}\left(|y|^2 + |z|^2\right) = \left(1 + \tfrac{1}{r_d^2}\,|y|^2\right)\left(1 + \tfrac{|z|^2}{r_d^2 + |y|^2}\right)$$

and, as a consequence,

$$\begin{aligned}
&\lim_{d\to+\infty}\left(1 + \tfrac{1}{r_d^2}\,|y|^2\right)^{-\frac{N+d}{2}} = e^{-\pi\,|y|^2}\,,\\
&\lim_{d\to+\infty}\textstyle\int_{\mathbb{R}^d} |v(y)|^2\,d\boldsymbol{\mu}_d = \int_{\mathbb{R}^N} |v|^2\,d\gamma\,,\\
&\lim_{d\to+\infty}\textstyle\int_{\mathbb{R}^d} |\nabla v|^2\left(1 + \tfrac{1}{r_d^2}\,|x|^2\right)^2 d\boldsymbol{\mu}_d = 4\int_{\mathbb{R}^N} |\nabla v|^2\,d\gamma\,,
\end{aligned}$$

where $d\gamma(y) = e^{-\pi\,|y|^2}\,dy$. However, the function w_v^d depends on d and the main difficulty is to obtain enough estimates on the parameters a_d, b_d and c_d to pass to the limit after integrating in all integrals with respect to z, which completes the proof of (7). See [28, Section 4] for further details. □

5 Stability of the Logarithmic Sobolev Inequality: A New Proof of Theorem 2

As explained in the Introduction and in Sect. 4, in [28] we proved Theorem 2 as a corollary of Theorem 1. In this section we give a new and complete proof of Theorem 2 that is independent of Theorem 1.

This gives us the opportunity to present some of the details both of the infinite dimensional limit mentioned in Sect. 4 and of the combination of continuous symmetrization and discrete competing symmetries flows mentioned in Sect. 3. In both cases the arguments in the present section are slightly simpler, but illustrate well the main underlying ideas. We also emphasize that, while it may have seemed that the latter argument relies on the conformal invariance of the Sobolev inequality (which enters through the operator U in the competing symmetries argument), this is not the case: in the setting of log-Sobolev (LSI), where there is no such conformal symmetry, there is another operator that serves the same purpose.

Just like in Sect. 3, we prove the quantitative version of the sharp logarithmic Sobolev inequality (Theorem 2) in two steps, one close and one far from the set of optimizers. Let us start with the following result that replaces Theorem 5 here. In fact, it is a consequence of Theorem 5.

Theorem 8 *There are explicit constants $\eta > 0$ and $\delta \in (0, 1/2)$ such that for all $N \in \mathbb{N}$ and for all nonnegative $u = 1 + r \in \mathrm{H}^1(\mathbb{S}^d)$ satisfying*

$$\int_{\mathbb{R}^N} r^2\,d\gamma \le \frac{\delta}{1-\delta} \tag{13}$$

and

$$\int_{\mathbb{R}^N} r\,d\gamma = 0 = \int_{\mathbb{R}^N} x_j\,r\,d\gamma\,, \quad j = 1, 2, \ldots, N\,, \tag{14}$$

one has

$$\int_{\mathbb{R}^N} |\nabla u|^2\,d\gamma - \pi \int_{\mathbb{R}^N} |u|^2 \ln\left(\frac{|u|^2}{\|u\|^2_{\mathrm{L}^2(\gamma)}}\right) d\gamma \ge \eta \int_{\mathbb{R}^N} r^2\,d\gamma\,.$$

The constant δ coincides with the corresponding constant in Theorem 5 and $\eta = 2\pi\,\epsilon_0$.

Proof Notice that $x \in \mathrm{L}^2(\gamma)$, so the orthogonality constraints raise no integration issues. We denote $\Sigma_d := \{x \in \mathbb{R}^{d+1} : |x| = \rho_d\}$ with $\rho_d := \sqrt{d/(2\pi)}$. The factor of $1/(2\pi)$ in the definition of ρ_d is necessary to get the π in the exponent of the Gaussian density. We integrate on Σ_d with respect to the uniform probability measure $d\mu_d$. Here, with a straightforward abuse, we use the same notation for the uniform probability measure on Σ_d as on the unit sphere. By rescaling our result in

Theorem 5 we find that

$$\int_{\Sigma_d} |\nabla R|^2 \, d\mu_d - \pi \, \frac{d-2}{2} \left(\left(\int_{\Sigma_d} (1+R)^{\frac{2d}{d-2}} \, d\mu_d \right)^{\frac{d-2}{d}} - \int_{\Sigma_d} (1+R)^2 \, d\mu_d \right)$$
$$\geq 2\pi \, \epsilon_0 \int_{\Sigma_d} \left(\frac{1}{\pi} \frac{2}{d-2} |\nabla R|^2 + R^2 \right) d\mu_d \, . \qquad (15)$$

This inequality is valid for all $R \in \mathrm{H}^1(\Sigma_d)$ such that

$$\left(\int_{\Sigma_d} R^{\frac{2d}{d-2}} \, d\mu_d \right)^{\frac{d-2}{d}} \leq \frac{\delta}{1-\delta} \qquad (16)$$

and

$$\int_{\Sigma_d} R \, d\mu_d = 0 = \int_{\Sigma_d} x_j \, R \, d\mu_d \, , \quad j = 1, \ldots, d+1 \, . \qquad (17)$$

Given a function $r \in \mathrm{H}^1(\gamma)$ and $d > N$, we apply this inequality to the function

$$R_d(x) := r(x_1, \ldots, x_N) - \int_{\Sigma_d} r \, d\mu_d - 2\pi \, \frac{d+1}{d} \sum_{n=1}^{N} x_n \int_{\Sigma_d} y_n \, r(y_1, \ldots, y_N) \, d\mu_d(y)$$

for $x \in \Sigma_d$. This function satisfies the orthogonality conditions (17). Note here that the functions $\sqrt{2\pi} \, \sqrt{(d+1)/d} \, x_j$ are L^2-normalized on Σ_d.

We now use the well-known fact that, as $d \to +\infty$, the marginal of $d\mu_d$ corresponding to the first N coordinates converges to $d\gamma$. Thus,

$$\lim_{d \to +\infty} \int_{\Sigma_d} |\nabla r|^2 \, d\mu_d = \int_{\mathbb{R}^N} |\nabla r|^2 \, d\gamma \, ,$$
$$\lim_{d \to +\infty} \int_{\Sigma_d} r^2 \, d\mu_d = \int_{\mathbb{R}^N} r^2 \, d\gamma \, ,$$
$$\lim_{d \to +\infty} \int_{\Sigma_d} r \, d\mu_d = \int_{\mathbb{R}^N} r \, d\gamma = 0 \, ,$$
$$\lim_{d \to +\infty} \int_{\Sigma_d} y_n \, r(y_1, \ldots, y_N) \, d\mu_d(y) = \int_{\mathbb{R}^N} y_n \, r \, d\gamma = 0 \, .$$

From this we conclude easily that

$$\lim_{d \to +\infty} \int_{\Sigma_d} |\nabla R_d|^2 \, d\mu_d = \int_{\mathbb{R}^N} |\nabla r|^2 \, d\gamma \, , \quad \lim_{d \to +\infty} \int_{\Sigma_d} R_d^2 \, d\mu_d = \int_{\mathbb{R}^N} r^2 \, d\gamma \, .$$

With some modest amount of effort one also finds that

$$\lim_{d\to+\infty}\int_{\Sigma_d} R_d^{\frac{2d}{d-2}}\,d\mu_d=\int_{\mathbb{R}^N} r^2\,d\gamma\,.$$

Assuming that the inequality in (13) is strict, the same is true for the left side when d is sufficiently large, and consequently the smallness condition (16) holds when d is sufficiently large. Thus, inequality (15) is valid for all sufficiently large d. The equality case in (13) can be obtained at the very end by a simple approximation argument.

Now, we drop the gradient term in the right side and letting $d\to+\infty$ we infer that

$$\int_{\mathbb{R}^N}|\nabla r|^2\,d\gamma-\pi\limsup_{d\to+\infty}\frac{d-2}{2}\left(\left(\int_{\Sigma_d}(1+R_d)^{\frac{2d}{d-2}}\,d\mu_d\right)^{\frac{d-2}{d}}-\int_{\Sigma_d}(1+R_d)^2\,d\mu_d\right)$$
$$\geq 2\pi\,\epsilon_0\int_{\mathbb{R}^N} r^2\,d\gamma\,.$$

Finally, we verify that

$$\limsup_{d\to+\infty}\frac{d-2}{2}\left(\left(\int_{\Sigma_d}(1+R_d)^{\frac{2d}{d-2}}\,d\mu_d\right)^{\frac{d-2}{d}}-\int_{\Sigma_d}(1+R_d)^2\,d\mu_d\right)$$
$$=\int_{\mathbb{R}^N}(1+r)^2\,\ln\left(\frac{(1+r)^2}{\|1+r\|^2_{\mathrm{L}^2(\gamma)}}\right)d\gamma\,.$$

In fact, if the orthogonality conditions were not present and the marginals would already be equal to their limit, this would follow from the fact that

$$\lim_{p\to1^+}\frac{1}{p-1}\left(\left(\int_{\mathbb{R}^N}h^p\,d\gamma\right)^{1/p}-\int_{\mathbb{R}^N}h\,d\gamma\right)=\int_{\mathbb{R}^N}h\ln\left(\frac{h}{\int_{\mathbb{R}^N}h\,d\gamma}\right)d\gamma\,,$$

valid on any measure space for any nonnegative function h that satisfies $h\in\mathrm{L}^1\cap\mathrm{L}^{p_0}(\gamma)$ for some $p_0>1$. Proving the latter fact is simple, as well as including the effect of the orthogonality conditions and the convergence of the marginals, so we shall omit it. These remarks complete the proof Theorem 8. □

We emphasize that in the previous proof we did not use Theorem 1, but rather Theorem 5. In this way we avoid having to control the distance to the set of optimizers in the high-dimensional limit, which seems harder than verifying the orthogonality conditions.

Proof of Theorem 2 As in the proof of Theorem 1, we first prove the result for nonnegative functions and then extend it to sign changing solutions. Let us denote by $\kappa^{\rm pos}$ the stability constant in the stability inequality restricted to nonnegative functions.

Step 1. Let η and δ be as in Theorem 8. For $0 \le u \in \mathrm{H}^1(\gamma)$ we distinguish two cases.

- The first case is where

$$\inf_{a\in\mathbb{R}^N,\, c\in\mathbb{R}} \int_{\mathbb{R}^N} \left|u - c\, e^{a\cdot x}\right|^2 d\gamma \le \delta \int_{\mathbb{R}^N} |u|^2\, d\gamma\,.$$

The infimum on the left-hand side is attained at some $a \in \mathbb{R}^N$ and $c \in \mathbb{R}$, as can be checked by optimizing $\int_{\mathbb{R}^N} \left|v - c\, e^{|a|^2/(2\pi) - \pi\, |x - a/\pi|^2/2}\right|^2 dx$, where $v(x) := u(x)\, e^{-\pi\, |x|^2/2}$. Let

$$\tilde u(y) := e^{-\, y\cdot a - \frac{|a|^2}{2\pi}}\, u\big(y + \tfrac{a}{\pi}\big)\,.$$

Then, by a simple computation involving an integration by parts and a change of variables,

$$\int_{\mathbb{R}^N} |\nabla \tilde u|^2\, d\gamma - \pi \int_{\mathbb{R}^N} |\tilde u|^2 \ln\left(\frac{|\tilde u|^2}{\|\tilde u\|^2_{\mathrm{L}^2(\gamma)}}\right) d\gamma$$
$$= \int_{\mathbb{R}^N} |\nabla u|^2\, d\gamma - \pi \int_{\mathbb{R}^N} |u|^2 \ln\left(\frac{|u|^2}{\|u\|^2_{\mathrm{L}^2(\gamma)}}\right) d\gamma\,.$$

Therefore, the deficit of $\tilde u$ coincides with that of u, while the infimum for $\tilde u$ among all functions of the form (4) is attained at the constant $c\, \exp\big(|a|^2/(2\pi)\big)$. Finally, by multiplying $\tilde u$ with a constant, we may assume that this constant is equal to one. To summarize, we may assume without loss of generality that the infimum in the theorem is attained at $a = 0$ and $c = 1$.

Let us set $r := u - 1$. Then the minimality implies that r satisfies the orthogonality conditions (14). Moreover, we have

$$\int_{\mathbb{R}^N} r^2\, d\gamma \le \delta \int_{\mathbb{R}^N} |u|^2\, d\gamma = \delta \left(1 + \int_{\mathbb{R}^N} r^2\, d\gamma\right),$$

so the smallness condition (13) is satisfied and we can apply Theorem 8. This yields the inequality in the theorem with a stability constant η.

- Next, we consider the case where

$$\inf_{a\in\mathbb{R}^N,\, c\in\mathbb{R}} \int_{\mathbb{R}^N} (u - c\, e^{x\cdot a})^2\, d\gamma > \delta \int_{\mathbb{R}^N} |u|^2\, d\gamma\,.$$

We argue similarly as we did in Sect. 3 concerning the Sobolev inequality, but there are some differences in this case.

For $f \in \mathrm{L}^2(\gamma)$ we denote by Uf its Gaussian rearrangement, that is, the function on $\mathbb{R}^N$ whose superlevel sets have the form $\{x \in \mathbb{R}^N : x_1 < \mu\}$ for some $\mu \in \mathbb{R}$ and have the same γ-measure as the corresponding superlevel sets of f. Moreover, we denote

$$Vf := e^{\frac{\pi}{2}|x|^2} \mathcal{R}\left(e^{-\frac{\pi}{2}|x|^2} f\right),$$

where $\mathcal{R}$ is, as before, the Euclidean rearrangement. Then, as shown in [18, Theorem 4.1], for any $0 \le f \in \mathrm{L}^2(\gamma)$ one has

$$f_n := (VU)^n f \to \|f\|_{\mathrm{L}^2(\gamma)} \quad \text{in} \quad \mathrm{L}^2(\gamma).$$

Moreover, $\|f_n\|_{\mathrm{L}^2(\gamma)} = \|f\|_{\mathrm{L}^2(\gamma)}$ and

$$n \mapsto \int_{\mathbb{R}^N} |\nabla f_n|^2 \, d\gamma - \pi \int_{\mathbb{R}^N} f_n^2 \ln\left(\frac{f_n^2}{\|f_n\|_{\mathrm{L}^2(\gamma)}^2}\right) d\gamma$$

is nonincreasing. This is the analogue of Theorem 6 in the present case.

We apply this procedure to our function u, which we assume here to be nonnegative, and obtain a sequence of functions u_n with constant $\mathrm{L}^2(\gamma)$-norm. Moreover, since

$$\inf_{a,\,c} \|u_n - c\, e^{a\cdot x}\|_{\mathrm{L}^2(\gamma)} \le \left\|u_n - \|u\|_{\mathrm{L}^2(\gamma)}\right\|_{\mathrm{L}^2(\gamma)} \to 0 \quad \text{as} \quad n \to +\infty,$$

there is an $n_0 \in \mathbb{N}$ such that

$$\inf_{a,\,c} \|u_{n_0} - c\, e^{a\cdot x}\|_{\mathrm{L}^2(\gamma)}^2 \ge \delta\, \|u\|_{\mathrm{L}^2(\gamma)}^2 > \inf_{a,\,c} \|u_{n_0+1} - c\, e^{a\cdot x}\|_{\mathrm{L}^2(\gamma)}^2 .$$

This replaces Lemma 7. We have

$$\begin{aligned}
&\frac{\int_{\mathbb{R}^N} |\nabla u|^2 \, d\gamma - \pi \int_{\mathbb{R}^N} |u|^2 \ln\left(\frac{|u|^2}{\|u\|_{\mathrm{L}^2(\gamma)}^2}\right) d\gamma}{\inf_{a,\,c} \|u - c\, e^{a\cdot x}\|_{\mathrm{L}^2(\gamma)}^2} \\
&\quad \ge \frac{\int_{\mathbb{R}^N} |\nabla u|^2 \, d\gamma - \pi \int_{\mathbb{R}^N} |u|^2 \ln\left(\frac{|u|^2}{\|u\|_{\mathrm{L}^2(\gamma)}^2}\right) d\gamma}{\|u\|_{\mathrm{L}^2(\gamma)}^2} \\
&\quad \ge \frac{\int_{\mathbb{R}^N} |\nabla u_{n_0}|^2 \, d\gamma - \pi \int_{\mathbb{R}^N} u_{n_0}^2 \ln\left(\frac{u_{n_0}^2}{\|u_{n_0}\|_{\mathrm{L}^2(\gamma)}^2}\right) d\gamma}{\|u\|_{\mathrm{L}^2(\gamma)}^2} .
\end{aligned}$$

We now use a continuous rearrangement flow to connect u_{n_0} to u_{n_0+1}. More precisely, we a consider a family of functions $(\mathsf{u}_\tau)_{n_0\le\tau\le n_0+1}$, where $\mathsf{u}_{n_0} := U u_{n_0}$ and $\mathsf{u}_{n_0+1} := u_{n_0+1}$. We define u_τ as $e^{\frac{\pi}{2}|x|^2}$ times the continuous (Euclidean) rearrangement of $e^{-\frac{\pi}{2}|x|^2}\,U u_{n_0}$ at parameter τ. In the same way as in [28, Lemma 36], one sees that

$$\tau \mapsto \inf_{a,\,c} \|\mathsf{u}_\tau - c\,e^{a\cdot x}\|^2_{\mathrm{L}^2(\gamma)}$$

is continuous, and therefore there is a $\tau_0 \in [n_0, n_0+1)$ such that

$$\inf_{a,\,c} \|\mathsf{u}_{\tau_0} - c\,e^{a\cdot x}\|^2_{\mathrm{L}^2(\gamma)} = \delta\,\|u\|^2_{\mathrm{L}^2(\gamma)}\,.$$

It follows that

$$\begin{aligned}
&\frac{\int_{\mathbb{R}^N} |\nabla u_{n_0}|^2\,d\gamma - \pi\int_{\mathbb{R}^N} u_{n_0}^2 \ln\left(\frac{u_{n_0}^2}{\|u_{n_0}\|^2_{\mathrm{L}^2(\gamma)}}\right) d\gamma}{\|u\|^2_{\mathrm{L}^2(\gamma)}}\\
&\qquad\ge \frac{\int_{\mathbb{R}^N} |\nabla \mathsf{u}_{\tau_0}|^2\,d\gamma - \pi\int_{\mathbb{R}^N} \mathsf{u}_{\tau_0}^2 \ln\left(\frac{\mathsf{u}_{\tau_0}^2}{\|\mathsf{u}_{\tau_0}\|^2_{\mathrm{L}^2(\gamma)}}\right) d\gamma}{\|u\|^2_{\mathrm{L}^2(\gamma)}}\\
&\qquad\qquad= \delta\,\frac{\int_{\mathbb{R}^N} |\nabla \mathsf{u}_{\tau_0}|^2\,d\gamma - \pi\int_{\mathbb{R}^N} \mathsf{u}_{n_0}^2 \ln\left(\frac{\mathsf{u}_{\tau_0}^2}{\|\mathsf{u}_{\tau_0}\|^2_{\mathrm{L}^2(\gamma)}}\right) d\gamma}{\inf_{a,\,c}\|\mathsf{u}_{\tau_0} - c\,e^{a\cdot x}\|^2_{\mathrm{L}^2(\gamma)}}\,.
\end{aligned}$$

We can apply the result in the first case to the function u_{τ_0} and infer that the right side is larger or equal than $\kappa^{\mathrm{pos}} := \delta\,\eta$. This concludes the proof in the case of nonnegative functions.

Step 2. We now prove the theorem in the general case, that is, for sign-changing functions.

We shall use the notation

$$D(v) := \int_{\mathbb{R}^N} |\nabla v|^2\,d\gamma - \pi\int_{\mathbb{R}^N} v^2 \ln\left(\frac{v^2}{\|v\|^2_{\mathrm{L}^2(\gamma)}}\right) d\gamma \quad \text{for } v \in \mathrm{H}^1(\gamma)\,.$$

Let $u = u_+ - u_- \in \mathrm{H}^1(\gamma)$. By homogeneity we can assume $\|u\|_{\mathrm{L}^2(\gamma)} = 1$. Replacing u by $-\,u$ if necessary, we can also assume that

$$m := \|u_-\|^2_{\mathrm{L}^2(\gamma)} \in [0, \tfrac{1}{2}]\,.$$

Then

$$D(u) = D(u_+) + D(u_-) + \pi\, h(m),$$

with

$$h(p) := -\big(p \ln p + (1-p)\ln(1-p)\big)\,.$$

Since the function $p \mapsto h(p)$ is monotone increasing and concave on $[0, 1/2]$, it holds that

$$h(p) \geq (2 \ln 2)\, p \quad \forall\, p \in [0, \tfrac{1}{2}]\,.$$

Thus, with $\kappa^{\rm pos}$ denoting the constant from Step 1,

$$\begin{aligned} D(u) &\geq D(u_+) + (2\pi\ \ln 2)\, m \geq \kappa^{\rm pos} \inf_{a,\,c} \|u_+ - c\, e^{a\cdot x}\|^2_{\mathrm{L}^2(\gamma)} + (2\pi\ \ln 2)\, \|u_-\|^2_{\mathrm{L}^2(\gamma)} \\ &\geq \frac{1}{2} \min\left\{\kappa^{\rm pos},\, 2\pi\ \ln 2\right\} \inf_{a,\,c} \|u - c\, e^{a\cdot x}\|^2_{\mathrm{L}^2(\gamma)}\,. \end{aligned}$$

This proves the inequality in the general case, with $\kappa = \frac{1}{2}\min\left\{\kappa^{\rm pos},\, 2\pi\ \ln 2\right\}$. It is straightforward to verify that $\kappa = \beta\,\pi/2$, β being the constant in the statement of Theorem 1. This ends our second proof of Theorem 2. □

Acknowledgments Partial support through US National Science Foundation grants DMS-1954995 (R.L.F.) and DMS-2154340 (M.L.), as well as through the Deutsche Forschungsgemeinschaft (DFG, German Research Foundation) Germany's Excellence Strategy EXC-2111-390814868 (R.L.F.) and through TRR 352—Project-ID 470903074 (R.L.F.), the French National Research Agency (ANR) project *Conviviality* (ANR-23-CE40-0003, J.D.) and the European Research Council under the Grant Agreement No. 721675 (RSPDE) *Regularity and Stability in Partial Differential Equations* (A.F.).

The authors are grateful to an anonymous referee for a careful rereading which helped to improve the manuscript.

References

1. Aubin, T.: Équations différentielles non linéaires et problème de Yamabe concernant la courbure scalaire. J. Math. Pures Appl. (9) **55**(3), 269–296 (1976)
2. Bakry, D., Émery, M.: Hypercontractivité de semi-groupes de diffusion. C. R. Acad. Sci. Paris Sér. I Math. **299**(15), 775–778 (1984)
3. Bakry, D., Émery, M.: Diffusions hypercontractives. In: Séminaire de probabilités, XIX, 1983/84, volume 1123 of Lecture Notes in Math., pp. 177–206. Springer, Berlin (1985)

4. Bakry, D., Émery, M.: Inégalités de Sobolev pour un semi-groupe symétrique. C. R. Acad. Sci. Paris Sér. I Math. **301**(8), 411–413 (1985)
5. Beckner, W.: Sharp Sobolev inequalities on the sphere and the Moser-Trudinger inequality. Ann. Math. (2) **138**(1), 213–242 (1993)
6. Bianchi, G., Egnell, H.: A note on the Sobolev inequality. J. Funct. Anal. **100**(1), 18–24 (1991)
7. Bidaut-Véron, M.F., Véron, L.: Nonlinear elliptic equations on compact Riemannian manifolds and asymptotics of Emden equations. Invent. Math. **106**(3), 489–539 (1991)
8. Bonforte, M., Dolbeault, J., Nazaret, B., Simonov, N.: Stability in Gagliardo-Nirenberg-Sobolev inequalities: flows, regularity and the entropy method. Mem. Amer. Math. Soc., 308(1554):viii+166, 2025
9. Brezis, H., Lieb, E.H.: Sobolev inequalities with remainder terms. J. Funct. Anal. **62**(1), 73–86 (1985)
10. Brigati, G., Dolbeault, J., Simonov, N.: Logarithmic Sobolev and interpolation inequalities on the sphere: Constructive stability results. Ann. Inst. Henri Poincaré C Anal. Non linéaire, 1–33 (2023)
11. Brigati, G., Dolbeault, J., Simonov, N.: On Gaussian interpolation inequalities. C. R. Math. Acad. Sci. Paris **362**, 21–44 (2024)
12. Brigati, G., Dolbeault, J., Simonov, N.: Stability for the logarithmic Sobolev inequality. J. Funct. Anal. **287**(8):Paper No. 110562, 21, 2024
13. Brock, F.: Continuous Steiner-symmetrization. Math. Nachr. **172**, 25–48 (1995)
14. Brock, F.: Continuous rearrangement and symmetry of solutions of elliptic problems. Proc. Indian Acad. Sci. Math. Sci. **110**(2), 157–204 (2000)
15. Caffarelli, L.A., Gidas, B., Spruck, J.: Asymptotic symmetry and local behavior of semilinear elliptic equations with critical Sobolev growth. Comm. Pure Appl. Math. **42**(3), 271–297 (1989)
16. Carlen, E.A.: Superadditivity of Fisher's information and logarithmic Sobolev inequalities. J. Funct. Anal. **101**(1), 194–211 (1991)
17. Carlen, E.A., Figalli, A.: Stability for a GNS inequality and the log-HLS inequality, with application to the critical mass Keller-Segel equation. Duke Math. J. **162**(3), 579–625 (2013)
18. Carlen, E.A., Loss, M.: Extremals of functionals with competing symmetries. J. Funct. Anal. **88**(2), 437–456 (1990)
19. Chen, S., Frank, R.L., Weth, T.: Remainder terms in the fractional Sobolev inequality. Indiana Univ. Math. J. **62**(4), 1381–1397 (2013)
20. Chen, L., Lu, G., Tang, H.: Stability of Hardy-Littlewood-Sobolev inequalities with explicit lower bounds Adv. Math., 450:Paper No. 109778, 28, 2024 s
21. Christ, M.: A sharpened Riesz-Sobolev inequality. arXiv: 1706.02007 (2017)
22. Demange, J.: Improved Gagliardo-Nirenberg-Sobolev inequalities on manifolds with positive curvature. J. Funct. Anal. **254**(3), 593–611 (2008)
23. Dolbeault, J., Esteban, M.J.: Improved interpolation inequalities and stability. Adv. Nonlinear Stud. **20**(2), 277–291 (2020)
24. Dolbeault, J., Toscani, G.: Stability results for logarithmic Sobolev and Gagliardo-Nirenberg inequalities. Int. Math. Res. Not. IMRN **2016**(2), 473–498 (2016)
25. Dolbeault, J., Esteban, M.J., Kowalczyk, M., Loss, M.: Improved interpolation inequalities on the sphere. Discrete Contin. Dyn. Syst. Ser. S **7**(4), 695–724 (2014)
26. Dolbeault, J., Esteban, M.J., Loss, M.: Nonlinear flows and rigidity results on compact manifolds. J. Funct. Anal. **267**(5), 1338–1363 (2014)
27. Dolbeault, J., Esteban, M.J., Loss, M.: Interpolation inequalities on the sphere: linear vs. nonlinear flows. Ann. Fac. Sci. Toulouse Math. (6) **26**(2), 351–379 (2017)
28. Dolbeault, J., Esteban, M.J., Figalli, A., Frank, R.L., Loss, M.: Sharp stability for Sobolev and log-Sobolev inequalities, with optimal dimensional dependence. Camb. J. Math. **13**(2), 359–430 (2025)
29. Eldan, R., Lehec, J., Shenfeld, Y.: Stability of the logarithmic Sobolev inequality via the Föllmer process. Ann. Inst. Henri Poincaré Probab. Stat. **56**(3), 2253–2269 (2020)

30. Fathi, M., Indrei, E., Ledoux, M.: Quantitative logarithmic Sobolev inequalities and stability estimates. Discrete Contin. Dyn. Syst. **36**(12), 6835–6853 (2016)
31. Frank, R.L.: Degenerate stability of some Sobolev inequalities. Ann. Inst. Henri Poincaré C Anal. Non linéaire **39**(6), 1459–1484 (2022)
32. Gidas, B., Ni, W.M., Nirenberg, L.: Symmetry of positive solutions of nonlinear elliptic equations in $\mathbb{R}^n$. In: Mathematical Analysis and Applications, Part A, volume 7 of Adv. in Math. Suppl. Stud., pp. 369–402. Academic Press, New York (1981)
33. Gozlan, N.: The deficit in the Gaussian log-Sobolev inequality and inverse Santaló inequalities. Int. Math. Res. Not. IMRN (17), 13396–13446 (2022)
34. Indrei, E., Kim, D.: Deficit estimates for the logarithmic Sobolev inequality. Differ. Integral Equations **34**(7–8), 437–466 (2021)
35. König, T.: On the sharp constant in the Bianchi–Egnell stability inequality. Bull. Lond. Math. Soc. **55**(4), 2070–2075 (2023)
36. König, T.: Stability for the Sobolev inequality: existence of a minimizer. J. Eur. Math. Soc. janvier 2025 (online first) https://doi.org/10.4171/jems/1582
37. Lieb, E.H.: Sharp constants in the Hardy-Littlewood-Sobolev and related inequalities. Ann. Math. (2) **118**(2), 349–374 (1983)
38. Lions, P.-L.: The concentration-compactness principle in the calculus of variations. The limit case. I. Rev. Mat. Iberoam. **1**(1), 145–201 (1985)
39. McKean, H.P.: Geometry of differential space. Ann. Probab. **1**, 197–206 (1973)
40. Rodemich, E.R.: The Sobolev inequalities with best possible constants. In: Analysis Seminar Caltech (1966)
41. Rosen, G.: Minimum value for c in the Sobolev inequality $\|\phi^3\| \leq c\,\|\nabla\phi\|^3$. SIAM J. Appl. Math. **21**, 30–32 (1971)
42. Talenti, G.: Best constant in Sobolev inequality. Ann. Mat. Pura Appl. (4) **110**, 353–372 (1976)

Orthonormal Strichartz Estimates for the Wave Equation and Related Geometric Inequalities

Neal Bez, Sanghyuk Lee, and Shohei Nakamura

Abstract For the one-sided wave propagator, we consider Strichartz estimates for orthonormal families of data. In three spatial dimensions, we provide some new progress towards what appears to be the natural conjecture regarding such estimates. In addition, we discuss connections with a spherical velocity averaging operator whose adjoint is a restricted X-ray transform along light rays.

Keywords Strichartz estimates · Orthonormal data · Wave equation · X-ray transform

1 Introduction

1.1 Some Broad Brushstrokes

At the heart of what we discuss in this article are estimates for solutions of time-evolution equations of the form

This work was supported by JSPS Kakenhi grant numbers 22H00098, 23K25777, 24H00024 (Bez), NRF grant RS-2024-00342160 (Lee), JSPS Overseas Research Fellowship and JSPS Kakenhi grant numbers 19K03546, 19H01796 and 21K13806 (Nakamura).

N. Bez (✉)
Graduate School of Mathematics, Nagoya University, Nagoya, Japan

Department of Mathematics, Graduate School of Science and Engineering, Saitama University, Saitama, Japan
e-mail: bez@math.nagoya-u.ac.jp

S. Lee
Department of Mathematical Sciences and RIM, Seoul National University, Seoul, South Korea
e-mail: shklee@snu.ac.kr

S. Nakamura
School of Mathematics, University of Birmingham, Birmingham, United Kingdom
e-mail: s.nakamura@bham.ac.uk

V. S. Georgiev, T. Ozawa (eds.), *Various Methods for the Analysis of PDEs*,
ICIAM2023 Springer Series 5, https://doi.org/10.1007/978-981-96-9164-7_4

$$i\partial_t\gamma - [D^\alpha, \gamma] = 0, \quad \gamma(0) = \gamma_0 \tag{1.1}$$

on $\mathbb{R}^d \times \mathbb{R}$ and for appropriate classes of operators γ. Here, $[\cdot, \cdot]$ denotes the commutator, $D = (-\Delta)^{1/2}$ and α is a real number. Our initial discussion will involve different values of α but our novel contributions in this present article are specific to the case $\alpha = 1$ corresponding to the wave equation.

Equation (1.1) is the interaction-free version of

$$i\partial_t\gamma - [D^\alpha, \gamma] = [w * \rho_\gamma, \gamma], \quad \gamma(0) = \gamma_0 \tag{1.2}$$

where ρ_γ is the density function associated with γ. If γ has integral kernel K defined on $\mathbb{R}^d \times \mathbb{R}^d$, then ρ_γ is (at least formally) given by the trace of the kernel on the diagonal; i.e. $\rho_\gamma(x) = K(x, x)$. For example, if γ_0 is finite-rank with a spectral decomposition

$$\gamma_0 = \sum_{j=1}^{N} \lambda_j \Pi_{f_j}$$

for orthonormal functions $f_1, \ldots, f_N \in L^2(\mathbb{R}^d)$, then

$$\rho_{\gamma_0}(x) = \sum_{j=1}^{N} \lambda_j |f_j(x)|^2.$$

In the above we write Π_f for the orthogonal projection given by $\Pi_f g = \langle g, f\rangle_{L^2} f$. Since the solution to (1.1) is given by

$$\gamma(t) = e^{-itD^\alpha}\gamma_0 e^{itD^\alpha}$$

we obtain

$$\rho_{\gamma(t)}(x) = \sum_{j=1}^{N} \lambda_j |e^{-itD^\alpha} f_j(x)|^2$$

in the free case $w = 0$. In particular, under the identification $f \leftrightarrow \Pi_f$, we see that the single-function equation

$$i\partial_t u - D^\alpha u = (w * |u|^2)u \tag{1.3}$$

corresponds to (1.2) with $N = 1$. This is already a wide framework; with $\alpha = 2$, the cubic nonlinear Schrödinger equation arises by taking w as the Dirac delta function at the origin, and Hartree-type nonlinear Schrödinger equations by taking $w(x) = |x|^{-\mu}$; see, for example, [10] for further background. We also refer the reader to, for example, [16], [17], [22] and [24] for some recent work on (1.3) pertaining to different values of α.

The operator-theoretic framework (1.2) is convenient for developing a rigorous theory of the dynamics of a system of infinitely many fermionic particles. It is apparent from the structure of (1.2) that obtaining control on the size of the density ρ_γ should play a key role in such a theory. For example, the Strichartz (space-time) estimates in this context take the form[1]

$$\Big\| \sum_j \lambda_j |e^{itD^\alpha} f_j|^2 \Big\|_{L_t^{\frac{q}{2}} L_x^{\frac{r}{2}}} \lesssim \|(\lambda_j)_j\|_{\ell^\beta} \tag{1.4}$$

for orthonormal families $(f_j)_j$ in the homogeneous Sobolev space $\dot{H}^s(\mathbb{R}^d)$, $s = \frac{d}{2} - \frac{d}{r} - \frac{\alpha}{q}$, and scalar sequences $(\lambda_j)_j$. Equivalently,

$$\Big\| \Big(\sum_j |e^{itD^\alpha} g_j|^2 \Big)^{\frac{1}{2}} \Big\|_{L_t^q L_x^r} \lesssim \Big(\sum_j \|g_j\|_{\dot{H}^s}^{2\beta} \Big)^{\frac{1}{2\beta}}$$

for families of orthogonal functions $(g_j)_j$ in $\dot{H}^s(\mathbb{R}^d)$. Frank–Lewin–Lieb–Seiringer [21] initiated the study of such estimates in the case $\alpha = 2$, and this has influenced a number of subsequent papers. Some of these papers have extended the framework for estimates of the form (1.4) to hold, and others are primarily focused on applying such estimates to develop a rigorous mathematical theory of equations like (1.2); see, for example, [5–9, 14, 15, 19, 20, 26–31, 34, 35, 38, 39]. Although the exponents s, q, r are forced to be linked by the scaling relation, the parameter β is unconstrained by such invariance and, put simply, the game is to establish such estimates with $\beta \geq 1$ as big as possible. The case $\beta = 1$ is equivalent to the classical Strichartz estimate

$$\|e^{itD^\alpha} g\|_{L_t^q L_x^r} \lesssim \|g\|_{\dot{H}^s} \tag{1.5}$$

and for this equivalence the orthogonality of the $(f_j)_j$ is redundant. In other words, the goal is to capture the gain from the orthogonality assumption in a quantitative manner via the size of the summability index β. In the case $\alpha \neq \{0, 1\}$, except certain endpoint/critical cases, the sharp value of β has been identified thanks to contributions in [9, 19–21].

As far as we are aware, *time-space* estimates

$$\Big\| \sum_j \lambda_j |e^{itD^\alpha} f_j|^2 \Big\|_{L_x^{\frac{r}{2}} L_t^{\frac{q}{2}}} \lesssim \|(\lambda_j)_j\|_{\ell^\beta}$$

[1] We use $A \lesssim B$ to mean $A \leq cB$ for an appropriate constant c.

for orthonormal families of initial data have received much less attention. With $q = \infty$ such estimates, even with localisation in both x and t variables on the left-hand side, give rise to almost-everywhere pointwise convergence of the density functions

$$\lim_{t \to 0} \rho_{\gamma(t)}(x) = \rho_{\gamma_0}(x)$$

for the solution of (1.1) with initial data γ_0 in certain Sobolev-type Schatten spaces (that is, $D^s \gamma_0 D^s$ belongs to the Schatten space $\mathcal{C}^\beta$ for certain β). This extension of Carleson's pointwise convergence problem for the free Schrödinger equation was first studied in [8], and we refer the reader to [7] for further background and some recent progress in this line of research.

1.2 The Wave Equation

Within the framework (1.4), the case $\alpha = 1$ corresponding to the one-sided wave propagator has offered the most resistance and the main goal of this paper is to present some new progress and observations in this direction. From now on we write[2]

$$Uf(x,t) := e^{itD} f(x) \simeq \int_{\mathbb{R}^d} e^{i(x \cdot \xi + t|\xi|)} \widehat{f}(\xi)\, d\xi.$$

Although ultimately one would wish to have a complete understanding of the estimate

$$\Big\| \sum_j \lambda_j |Uf_j|^2 \Big\|_{L_t^{\frac{q}{2}} L_x^{\frac{r}{2}}} \lesssim \|(\lambda_j)_j\|_{\ell^\beta} \tag{1.6}$$

for orthonormal families $(f_j)_j$ in $\dot{H}^s(\mathbb{R}^d)$, $s = \frac{d}{2} - \frac{d}{r} - \frac{1}{q}$, here we make the situation slightly nicer by introducing the following frequency-localised version

$$\Big\| \sum_j \lambda_j |U_0 f_j|^2 \Big\|_{L_t^{\frac{q}{2}} L_x^{\frac{r}{2}}} \lesssim \|(\lambda_j)_j\|_{\ell^\beta} \tag{1.7}$$

[2] The notation $A \simeq B$ means $A = cB$ for an appropriate constant c. In other words, this notation is used to hide unimportant constants in the case of an identity, and so A, B and c may be complex numbers.

for families of orthonormal functions $(f_j)_j$ in $L^2(\mathbb{R}^d)$. Here $U_0 f := U P_0 f$ and P_0 is the Littlewood–Paley operator given by

$$\widehat{P_0 f}(\xi) = \chi(|\xi|)\widehat{f}(\xi),$$

where $\chi \in C_c^\infty(\mathbb{R})$ is supported in $[\frac{1}{2}, 2]$. As far as we are aware, the orthogonality constraint means that, strictly speaking, (1.6) does not immediately imply (1.7), but it is difficult to imagine a proof of (1.6) which does not equally well apply to the frequency-localised operator U_0 too. As we shall see, though, the presence of the frequency localisation certainly helps when analysing certain oscillatory integrals which naturally arise in this context and in this sense we think of (1.7) as a more manageable estimate. On the other hand, (1.7) can often be upgraded to a certain restricted weak-type version of (1.6) (see, for example, [6, Lemma 3.1]).

Our focus here is on the sharp summability exponent β and, given where things currently stand in the existing literature, we are prepared to accept the slightly weaker version in (1.7). Indeed, we shall also put forward some evidence to suggest that obtaining the sharp value of β may be quite challenging even in the frequency-localised framework (1.7).

To state the natural conjecture for (1.7), first we introduce the notation $\beta_\kappa(q, r)$ via the relation

$$\frac{1}{\beta_\kappa(q,r)} = \frac{1}{\kappa q} + \frac{2}{r}.$$

Here, $\kappa > 0$ and note that $\beta_\kappa(q, r)$ is clearly monotone increasing with respect to κ for fixed (q, r). Since (1.7) is stronger than the single-function (frequency-localised) Strichartz estimate

$$\|U_0 g\|_{L_t^q L_x^r} \lesssim \|g\|_{L^2} \tag{1.8}$$

it is necessary to restrict attention to (q, r) for which

$$q, r \in [2, \infty], \quad \frac{1}{q} \leq \frac{d-1}{2}\left(\frac{1}{2} - \frac{1}{r}\right).$$

Also, one can show that for (1.7) to hold, we have the following necessary condition

$$\beta \leq \min\left\{\beta_{\frac{d}{2}}(q, r), \frac{q}{2}\right\} =: \beta_*(q, r). \tag{1.9}$$

Indeed, it was shown in [6, 9] that (1.9) is necessary for (1.6) and a quick inspection of the proof reveals that the data used to generate the two upper bounds are frequency-localised to an annulus. As a result, in particular this rules out the possibility of obtaining anything beyond the case $\beta = 1$ when $q = 2$.

When $q = \infty$, the estimate (1.7) is valid for all $r \in [2, \infty]$ and[3] $\beta = \beta_\kappa(\infty, r) = \frac{r}{2}$. To prove such estimates, it suffices by interpolation to consider the two endpoint cases $r = 2$ and $r = \infty$. Since $\beta_\kappa(\infty, 2) = 1$, the claim when $r = 2$ easily follows from conservation of energy and the triangle inequality. When $r = \infty$, clearly one may reduce to the case $\|(\lambda_j)_j\|_\infty = 1$ and, following the argument in [5], Bessel's inequality implies

$$\sum_j |U_0 f_j(x,t)|^2 \simeq \sum_j |\langle \chi_{x,t}, \widehat{f_j}\rangle_{L^2}|^2 \lesssim \|\chi_{x,t}\|_{L^2}^2 \lesssim 1$$

uniformly in (x, t), where $\chi_{x,t}(\xi) := e^{-i(x\cdot\xi + t|\xi|)}\chi(|\xi|)$. Of course, we are using the orthogonality of the $(\widehat{f_j})_j$ which is inherited from the orthogonality of the $(f_j)_j$ thanks to Parseval's identity.

Remark Interestingly, (1.6) fails when $q = \infty$, $r \in (2, \infty]$ and $\beta = \frac{r}{2}$, but is valid for $\beta < \frac{r}{2}$ thanks to Lieb's orthonormal Sobolev inequality [36].

For other values of q, the following appears to be a reasonable conjecture (although the case $q = \infty$ is settled for reasons explained above, it is convenient to include this case in the following conjecture). Without the frequency-localisation, the same conjecture explicitly appears in [6].

Conjecture 1.1 *Let $d \geq 2$, $q \in (2, \infty]$, $r \in [2, \infty]$, and $\frac{1}{q} \leq \frac{d-1}{2}(\frac{1}{2} - \frac{1}{r})$. Then the estimate (1.7) holds in each of the following cases.*

(1) $\beta \leq \beta_{\frac{d}{2}}(q, r)$ *when* $\frac{d}{r} > \frac{d-1}{q}$.
(2) $\beta < \frac{q}{2}$ *when* $\frac{d}{r} \leq \frac{d-1}{q}$.

If we ignore the issue of what is happening in the case $\beta = \beta_*(q, r)$, the conjecture says that $\beta_*(q, r)$ is the sharp exponent in all admissible cases. Some progress on the conjecture has been made thanks to [6, 9, 20] and we refer the reader to [6] for a detailed account of prior results in this direction. Here we emphasise that it is known that (1.7) holds for $\beta \leq \beta_{\frac{d-1}{2}}(q, r)$ when $\frac{d-1}{r} > \frac{d-2}{q}$ (see [9, 20]). Conjecture 1.1(1) posits that one may extend the range of β as far as $\beta_{\frac{d}{2}}(q, r)$ in the subregion where $\frac{d}{r} > \frac{d-1}{q}$. In other words, if we set

$$A = \left(\frac{d-2}{2d}, \frac{d-1}{2d}\right), \quad B = \left(\frac{(d-1)^2}{2(d^2+1)}, \frac{d(d-1)}{2(d^2+1)}\right), \quad C = \left(\frac{1}{2}, 0\right)$$

then (1.7) for

$$(\tfrac{1}{r}, \tfrac{1}{q}) \in OAC \setminus (O, A], \quad \beta \leq \beta_{\frac{d-1}{2}}(q, r)$$

[3] When written in this way, there is no restriction on $\kappa > 0$ but, in light of the forthcoming Conjecture 1.1, we have the case $\kappa = \frac{d}{2}$ in mind.

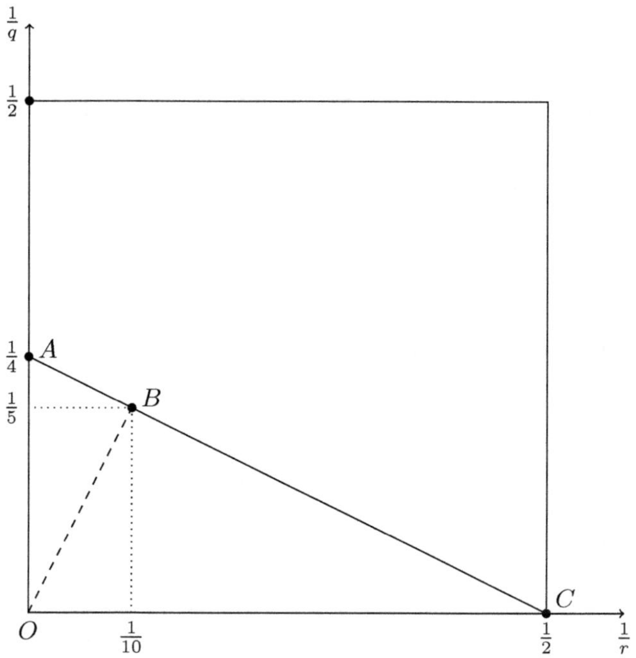

Fig. 1 $d = 2$

is known, and the pressing issue here is to obtain (1.7) for

$$(\tfrac{1}{r}, \tfrac{1}{q}) \in OBC \setminus (O, B], \quad \beta \le \beta_{\frac{d}{2}}(q, r).$$

See Figs. 1 and 2 for the cases $d = 2$ and $d = 3$, respectively. Here, XYZ denotes the triangle in the plane with vertices at X, Y and Z, and $(X, Y]$ and $[X, Y]$ are line segments connecting X and Y with the endpoint X removed in the former case.

If we set $D = (\frac{1}{6}, \frac{1}{4})$, then we partially verify the conjecture when $d = 3$ in the following sense.

Theorem 1.2 *Let $d = 3$. Then Conjecture 1.1 holds when*

$$(\tfrac{1}{r}, \tfrac{1}{q}) \in (ODC)^\circ \cup [O, C].$$

For other dimensions, we see no obstacle to extending Theorem 1.2 but the proof is particularly clean when $d = 3$ and so we restrict ourselves to this case in order not to obfuscate matters.

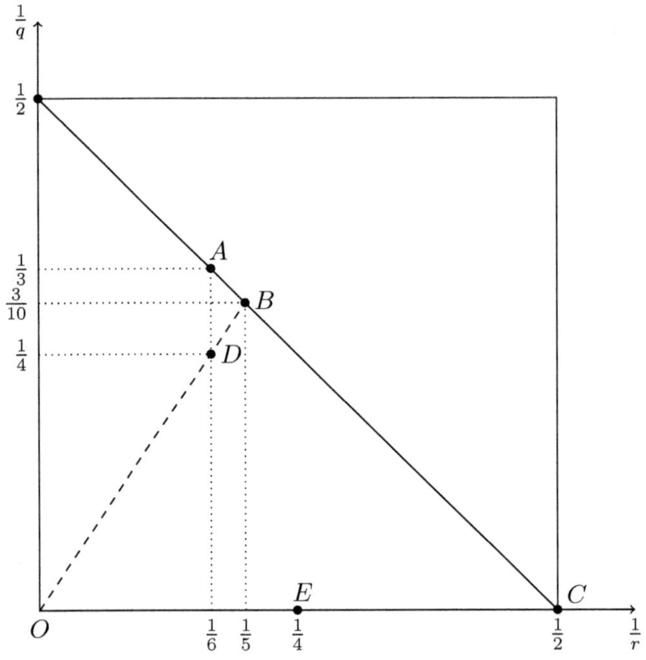

Fig. 2 $d = 3$

1.3 Kinetic Averages

Another way of gauging the difficulty of Conjecture 1.1 is to use a semi-classical limiting argument and consider the corresponding velocity averaging estimates which arise. If (1) in Conjecture 1.1 is true then from [9, Proposition 12] we know that the estimate

$$\|\rho f\|_{L_t^{\frac{q}{2}} L_x^{\frac{r}{2}}(\mathbb{R}\times\mathbb{R}^d)} \lesssim \|f\|_{L^{\beta}_{x,v}(\mathbb{R}^d\times\mathbb{R}^d)} \tag{1.10}$$

holds for all $f \in L^{\beta}_{x,v}$ provided

$$(\tfrac{1}{r}, \tfrac{1}{q}) \in OBC \setminus (O, B], \quad \beta = \beta_{\frac{d}{2}}(q, r).$$

Here, ρ is the velocity averaging operator given by

$$\rho f(x,t) := \int_{\mathbb{R}^d} f(x - tv', v)\chi(|v|)^2 \, dv$$

and $v' := |v|^{-1}v \in \mathbb{S}^{d-1}$. The adjoint operator ρ^* is given by

$$\rho^* g(x, v) := \chi(|v|)^2 \int_{\mathbb{R}} g(x + tv', t)\, dt$$

and therefore, by polar coordinates,

$$\|\rho^* g\|_{L^{\beta'}_{x,v}(\mathbb{R}^d \times \mathbb{R}^d)} \simeq \left\| \int_{\mathbb{R}} g(x + t\theta, t)\, dt \right\|_{L^{\beta'}_{x,\theta}(\mathbb{R}^d \times \mathbb{S}^{d-1})}.$$

Thus, (1.10) is equivalent to

$$\|\varrho^* g\|_{L^{\beta'}_{x,\theta}(\mathbb{R}^d \times \mathbb{S}^{d-1})} \lesssim \|g\|_{L_t^{\frac{\widetilde{q}}{2}} L_x^{\frac{\widetilde{r}}{2}}(\mathbb{R}^d \times \mathbb{R})} \tag{1.11}$$

where

$$\varrho^* g(x, \theta) := \int_{\mathbb{R}} g(x + t\theta, t)\, dt, \qquad (x, \theta) \in \mathbb{R}^d \times \mathbb{S}^{d-1}$$

and $\widetilde{q}$ denotes the "$\frac{1}{2}$-conjugate" of q given by

$$\frac{1}{q} + \frac{1}{\widetilde{q}} = \frac{1}{2}.$$

By duality once again, (1.10) is equivalent to

$$\|\varrho f\|_{L_t^{\frac{q}{2}} L_x^{\frac{r}{2}}(\mathbb{R}^d \times \mathbb{R})} \lesssim \|f\|_{L^{\beta}_{x,\theta}(\mathbb{R}^d \times \mathbb{S}^{d-1})}, \tag{1.12}$$

where

$$\varrho f(x, t) := \int_{\mathbb{S}^{d-1}} f(x - t\theta, \theta)\, d\sigma(\theta), \qquad (x, t) \in \mathbb{R}^d \times \mathbb{R}$$

and $d\sigma$ denotes the induced Lebesgue measure on $\mathbb{S}^{d-1}$. To summarise, we have the following.

Proposition 1.3 *If (1) in Conjecture 1.1 is true, then* (1.11) *and* (1.12) *hold whenever*

$$(\tfrac{1}{r}, \tfrac{1}{q}) \in OBC \setminus (O, B], \quad \beta = \beta_{\frac{d}{2}}(q, r).$$

The operator ϱ is the spherical velocity average of the solution to the kinetic transport equation. Estimate (1.12) in special case $\beta = 2$ is well understood and the full range of estimates can be found in the forthcoming Corollary 3.2. In fact, when $\beta = 2$, estimates for ϱ in the wider framework of hyperbolic Sobolev spaces were proved in [2, 4, 11, 12]. These estimates take the form

$$\|D_+^{s_+} D_-^{s_-} \varrho f\|_{L_t^a L_x^b(\mathbb{R}^d \times \mathbb{R})} \lesssim \|f\|_{L^2_{x,\theta}(\mathbb{R}^d \times \mathbb{S}^{d-1})}$$

and we expand on this discussion in Sect. 3. On the other hand, obtaining a full characterisation of the exponents (q, r, β) for which (1.12) holds seems to be a very challenging problem.

The adjoint operator ϱ^* is an X-ray like transform except that the directions over which the line integrals are taken are restricted to light rays

$$\{(\theta, 1) : \theta \in \mathbb{S}^{d-1}\}.$$

Wolff [43] considered mixed-norm estimates for such an operator but, in contrast to (1.11), the mixed-norm appears on the left-hand side of the estimate. Nevertheless, there is of course an overlap of the two frameworks when $q = r$ and, given the complexity of Wolff's arguments, it is tempting to think that this adds further weight to the impression that Conjecture 1.1 is a challenging problem. We develop this discussion further in Sect. 4.

Remark When the velocity variable runs over all of $\mathbb{R}^d$, $d \geq 1$, the analogous family of estimates to (1.10) go back to the work of Castella–Perthame [13] and take the form

$$\left\| \int_{\mathbb{R}^d} f(x - tv, v)\, dv \right\|_{L_t^{\frac{q}{2}} L_x^{\frac{r}{2}}(\mathbb{R}^d \times \mathbb{R})} \lesssim \|f\|_{L^\beta_{x,v}(\mathbb{R}^d \times \mathbb{R}^d)}. \tag{1.13}$$

This family of estimates is very well understood and it is known that (1.13) holds if and only if

$$\frac{1}{q} = \frac{d}{2}\left(\frac{1}{2} - \frac{1}{r}\right), \quad r \in \left[2, \frac{2(d+1)}{d-1}\right).$$

The sufficiency of these conditions is due to [13, 32], and the failure of the endpoint case $r = \frac{2(d+1)}{d-1}$ was proved in [25, 40] when $d = 1$ and [1] when $d \geq 2$. The estimate (1.13) may also be obtained in all admissible cases via a semi-classical limit from the orthonormal Strichartz estimate (1.4) when $\alpha = 2$ (which were established in [20, 21]).

Organisation Theorem 1.2 is proved in Sect. 2. In Sect. 3, we give a proof of the corresponding kinetic estimates (including an endpoint case) by following the Fourier-analytic argument in [2]. Finally, in Sect. 4, we make a few observations regarding the restricted X-ray transform estimate (1.11).

2 Proof of Theorem 1.2

By interpolation with (1.7) at O and C with $\beta = \beta_{\frac{3}{2}}(q,r)$, it suffices to prove (1.7) with $\beta = \beta_{\frac{3}{2}}(q,r)$ at points $(\frac{1}{r}, \frac{1}{q})$ which approach D. In fact, we prove that this is the case whenever $(\frac{1}{r}, \frac{1}{q}) \in (D, E]$, where $E = (\frac{1}{4}, 0)$. As we shall see, the reason why this case is particularly susceptible is that $\beta_{\frac{3}{2}}(q,r) = 2$ for such exponents.

Theorem 2.1 *Let* $d = 3$,

$$\frac{1}{q} + \frac{3}{r} = \frac{3}{4}, \qquad r \in [4, 6), \tag{2.1}$$

and $\beta = 2$. *Then* (1.7) *holds.*

As a key ingredient in our proof of Theorem 2.1, we have the following classical bounds on the averaging operator $\mathcal{A}_t$ given by

$$\mathcal{A}_t f(x) := \int_{\mathbb{S}^{d-1}} f(x - t\theta)\, d\sigma(\theta).$$

Geometrically we think of this as an averaging operator associated with the sphere of radius t, but there is no problem allowing $t < 0$ too.

Proposition 2.2 *Let* $d \geq 2$. *Then*

$$\|\mathcal{A}_t f\|_{L^b(\mathbb{R}^d)} \lesssim |t|^{d(\frac{1}{b} - \frac{1}{a})} \|f\|_{L^a(\mathbb{R}^d)}$$

holds whenever $(\frac{1}{a}, \frac{1}{b})$ *lies in the closed triangle with vertices* $(0,0)$, $(1,1)$ *and* $(\frac{d}{d+1}, \frac{1}{d+1})$.

By a simple rescaling argument, the proof of Proposition 2.2 can be reduced to the case $t = 1$. The claimed bounds on $\mathcal{A}_1$ are well-known (see [37]).

Proof of Theorem 2.1 By the duality principle of Frank–Sabin [20], it suffices to prove

$$\Lambda := \int |W(x,t)|^2 |W(x',t')|^2 |K(x - x', t - t')|^2 \, dx dx' dt dt' \lesssim \|W\|^4_{L_t^{\widetilde{q}} L_x^{\widetilde{r}}} \tag{2.2}$$

under the restriction

$$\frac{1}{\widetilde{q}} + \frac{3}{\widetilde{r}} = \frac{5}{4}, \qquad \widetilde{r} \in (3, 4], \tag{2.3}$$

and where

$$K(x,t) := \int_{\mathbb{R}^3} e^{i(x\cdot\xi + t|\xi|)} \chi(|\xi|)^2 \, d\xi.$$

The integration in Λ is taken over $(x, x', t, t') \in \mathbb{R}^3 \times \mathbb{R}^3 \times \mathbb{R} \times \mathbb{R}$. In the course of the proof, we divide this into various regions Ω_j, and in each case we write

$$\Lambda_j := \int_{\Omega_j} |W(x,t)|^2 |W(x',t')|^2 |K(x-x',t-t')|^2 \, dxdx'dtdt'.$$

Since we are working in three spatial dimensions, we have the exact formula[4]

$$\widehat{d\sigma}(x) \simeq \frac{1}{|x|} \sin|x| = \frac{1}{2i|x|}(e^{i|x|} - e^{-i|x|}).$$

By setting $\chi_0(r) := r\chi(r)^2$, this leads to

$$\begin{aligned} K(x,t) &= \int_0^\infty e^{itr} r\chi_0(r) \widehat{d\sigma}(rx)\, dr \\ &\simeq \frac{1}{|x|}\int_0^\infty e^{ir(t+|x|)}\chi_0(r)\, dr - \frac{1}{|x|}\int_0^\infty e^{ir(t-|x|)}\chi_0(r)\, dr \\ &= \frac{1}{|x|}(\widehat{\chi_0}(t+|x|) - \widehat{\chi_0}(t-|x|)). \end{aligned}$$

□

Main Contribution As a result of the rapid decay of $\widehat{\chi_0}$, the main contribution to Λ arises when

$$||x-x'| - |t-t'|| \le 1.$$

First, we consider the region

$$\Omega_1 := \{(x,x',t,t') : ||x-x'| - |t-t'|| \le 1, |t-t'| \ge 100\}.$$

Then

$$\begin{aligned} \Lambda_1 &\lesssim \int_{\Omega_1} |W(x,t)|^2 |W(x',t')|^2 \frac{dxdx'}{|x-x'|^2}\, dtdt' \\ &= \int_{\Omega_1'} |W(|t-t'|y,t)|^2 |W(|t-t'|y',t')|^2 \frac{dydy'}{|y-y'|^2} |t-t'|^4 \, dtdt' \\ &\lesssim \int_{\Omega_1'} |W(|t-t'|y,t)|^2 |W(|t-t'|y',t')|^2 dydy' |t-t'|^4 \, dtdt' \end{aligned}$$

[4] Recall that $A \simeq B$ means $A = cB$ for an appropriate constant c.

where

$$\Omega_1' := \{(y, y', t, t') : ||y - y'| - 1| \leq \tfrac{1}{|t-t'|}, |t - t'| \geq 100\}.$$

Lemma 2.3 *Let* $p \in [\frac{4}{3}, 2]$ *and* $I \subseteq [-1, 1]$. *Then*

$$\|f * 1_I(R(1 - |\cdot|))\|_{L^{p'}(\mathbb{R}^3)} \lesssim \frac{1}{R}\|f\|_{L^p(\mathbb{R}^3)}$$

for any $R \geq 1$.

Proof Note

$$\int_{\mathbb{R}^3} \varphi(y) 1_J(|y|)\, dy = \int_J t^2 \int_{\mathbb{S}^2} \varphi(t\theta)\, d\sigma(\theta) dt.$$

This means we may write

$$f * 1_I(R(1 - |\cdot|))(x) = \int_J t^2 \mathcal{A}_t f(x)\, dt$$

where $J := 1 - \frac{1}{R} I$, and hence by Proposition 2.2 we obtain

$$\begin{aligned} \|f * 1_I(R(1 - |\cdot|))\|_{L^{p'}(\mathbb{R}^3)} &\leq \int_J t^2 \|\mathcal{A}_t f\|_{L^{p'}(\mathbb{R}^3)}\, dt \\ &\lesssim \|f\|_{L^p(\mathbb{R}^3)} \int_J t^{5-\frac{6}{p}}\, dt. \end{aligned}$$

Since $I \subseteq [-1, 1]$, we have $|J| \lesssim \frac{1}{R}$ and the claimed estimate follows. □

By Lemma 2.3, for $\widetilde{r} \in [\frac{8}{3}, 4]$ we have

$$\begin{aligned} \Lambda_1 &\lesssim \int_{|t-t'|\geq 100} \||W(|t - t'|y, t)|^2\|_{L_y^{\widetilde{r}/2}} \||W(|t - t'|y', t')|^2\|_{L_y^{\widetilde{r}/2}} |t - t'|^3\, dt dt' \\ &= \int_{|t-t'|\geq 100} \|W(\cdot, t)\|_{L^{\widetilde{r}}}^2 \|W(\cdot, t')\|_{L^{\widetilde{r}}}^2 \frac{dt dt'}{|t - t'|^{\frac{12}{\widetilde{r}}-3}} \end{aligned}$$

and by the Hardy–Littlewood–Sobolev inequality, under the condition (2.3), we have

$$\Lambda_1 \lesssim \|W\|_{L_t^{\widetilde{q}} L_x^{\widetilde{r}}}^4.$$

Also, for the region

$$\Omega_2 := \{(x,x',t,t') : ||x-x'|-|t-t'|| \le 1, |t-t'| \le 100\}$$

we have

$$\Lambda_2 \lesssim \int_{|t-t'|\le 100} \int_{|x-x'|\le 101} |W(x,t)|^2 |W(x',t')|^2 dx dx' \, dt dt'.$$

Hence, by Young's convolution inequality, we obtain

$$\Lambda_2 \lesssim \|W\|^4_{L_t^{\widetilde{q}} L_x^{\widetilde{r}}}$$

whenever $\widetilde{q}, \widetilde{r} \in [2,4]$.

Tail Terms It remains to handle the contribution to Λ from the region

$$||x-x'|-|t-t'|| \ge 1$$

for which we capitalise on the rapid decay of $\widehat{\chi}_0$. First, we consider

$$\begin{aligned}\Omega_3 &:= \{(x,x',t,t') : ||x-x'|-|t-t'|| \ge 1, |t-t'| \le \tfrac{1}{2}\} \\ &= \bigcup_{k=0}^{\infty} \{(x,x',t,t') : ||x-x'|-|t-t'|| \in J_k, |t-t'| \le \tfrac{1}{2}\} =: \bigcup_{k=0}^{\infty} \Omega_{3,k},\end{aligned}$$

where $J_k := [2^k, 2^{k+1})$. Then the rapid decay of $\widehat{\chi}_0$ implies

$$\Lambda_3 \lesssim_N \sum_{k=0}^{\infty} \frac{1}{2^{(N+2)k}} \int_{|t-t'|\lesssim 1} \int_{|x-x'|\lesssim 2^k} |W(x,t)|^2 |W(x',t')|^2 \, dx dx' dt dt'$$

and hence, by Young's convolution inequality

$$\Lambda_3 \lesssim_N \|W\|^4_{L_t^{\widetilde{q}} L_x^{\widetilde{r}}} \sum_{k=0}^{\infty} \frac{2^{\mu k}}{2^{(N+2)k}}$$

whenever $\widetilde{q}, \widetilde{r} \in [2,4]$, and where $\mu := 3(2-\frac{4}{\widetilde{r}})$. Clearly, a sufficiently large choice of N will guarantee that the above series is convergent.

Next, we consider

$$\Omega_4 := \{(x,x',t,t') : 1 \le ||x-x'|-|t-t'|| \le \tfrac{1}{2}|t-t'|, |t-t'| \ge \tfrac{1}{2}\}.$$

Then

$$\Lambda_4 \lesssim_N \int_{\Omega_4'} |W(|t-t'|y,t)|^2 |W(|t-t'|y',t')|^2 \frac{dydy'}{||y-y'|-1|^N} \frac{dtdt'}{|t-t'|^{N-4}}$$

where

$$\Omega_4' := \{(y,y',t,t') : \tfrac{1}{|t-t'|} \le ||y-y'|-1| \le \tfrac{1}{2}, |t-t'| \ge \tfrac{1}{2}\}.$$

Now, if $\delta \in (0,\frac{1}{2})$ and $j_0 \in \mathbb{N}$ is such that $\delta \in J_{-j_0-1}$, then

$$\begin{aligned}\int_{||y-y'|-1| \in [\delta,\frac{1}{2}]} & F(y)G(y') \frac{dydy'}{|y-y'|-1|^N} \\ &\lesssim \sum_{j=1}^{j_0} 2^{jN} \int_{||y-y'|-1| \in J_{-j-1}} F(y)G(y')\, dydy'\end{aligned}$$

and by Lemma 2.3, for $\widetilde{r} \in [\frac{8}{3},4]$, we have

$$\Lambda_4 \lesssim_N \int_{|t-t'| \ge \frac{1}{2}} \| |W(|t-t'|y,t)|^2 \|_{L_y^{\widetilde{r}/2}} \| |W(|t-t'|y',t')|^2 \|_{L_y^{\widetilde{r}/2}} |t-t'|^3 \, dtdt'.$$

As with Λ_1, it follows that $\Lambda_4 \lesssim \|W\|^4_{L_t^{\widetilde{q}} L_x^{\widetilde{r}}}$ holds under (2.3).

Next, we consider

$$\Omega_5 := \{(x,x',t,t') : ||x-x'|-|t-t'|| > |t-t'|, |t-t'| \ge \tfrac{1}{2}\}$$

Then

$$\begin{aligned}\Lambda_5 &\lesssim_N \int_{\Omega_5} |W(x,t)|^2 |W(x',t')|^2 \frac{dxdx'}{|x-x'|^2 ||x-x'|-|t-t'||^N} dtdt' \\ &= \int_{\Omega_5'} |W(|t-t'|y,t)|^2 |W(|t-t'|y',t')|^2 \frac{dydy'}{||y-y'|-1|^N} \frac{dtdt'}{|t-t'|^{N-4}} \\ &\lesssim \sum_{k=0}^{\infty} \frac{1}{2^{(k+2)N}} \int_{\Omega_{5,k}'} |W(|t-t'|y,t)|^2 |W(|t-t'|y',t')|^2 \, dydy' \frac{dtdt'}{|t-t'|^{N-4}}\end{aligned}$$

where

$$\begin{aligned}\Omega_5' &:= \{(y,y',t,t') : ||y-y'|-1| > 1, |t-t'| \ge \tfrac{1}{2}\} \\ &= \bigcup_{k=0}^{\infty} \{(y,y',t,t') : ||y-y'|-1| \in (2^k, 2^{k+1}], |t-t'| \ge \tfrac{1}{2}\}.\end{aligned}$$

Note that $||y-y'|-1| \in (2^k, 2^{k+1}]$ and $k \geq 0$ imply $|y-y'|-1 \in (2^k, 2^{k+1}]$ and so $|y-y'| \sim 2^k$. Thus, by Young's convolution inequality,

$$\Lambda_5 \lesssim \sum_{k=0}^{\infty} \frac{1}{2^{(N+2)k}} \int_{|t-t'|\gtrsim 1} \int_{|y-y'|\lesssim 2^k} |W(|t-t'|y,t)|^2 |W(|t-t'|y',t')|^2 \, dy dy' \times \frac{dtdt'}{|t-t'|^{N-4}}$$

$$\lesssim \sum_{k=0}^{\infty} \frac{2^{\mu k}}{2^{(N+2)k}} \int_{|t-t'|\gtrsim 1} \|W(\cdot,t)\|_{\widetilde{r}}^2 \|W(\cdot,t')\|_{\widetilde{r}}^2 \frac{dtdt'}{|t-t'|^{\frac{12}{\widetilde{r}}+N-4}}$$

whenever $\widetilde{r} \in [2,4]$, and where $\mu := 3(2-\frac{4}{\widetilde{r}})$. By choosing N sufficiently large, we obtain $\Lambda_5 \lesssim \|W\|^4_{L_t^{\widetilde{q}} L_x^{\widetilde{r}}}$ whenever $\widetilde{q}, \widetilde{r} \in [2,4]$.

It remains to consider

$$\begin{aligned}
\Omega_6 &:= \{(x,x',t,t') : \tfrac{1}{2}|t-t'| < ||x-x'|-|t-t'|| \leq |t-t'|, |t-t'| \geq \tfrac{1}{2}\} \\
&= \{(x,x',t,t') : \tfrac{1}{2}|x-x'| < |t-t'| \leq \tfrac{2}{3}|x-x'|, |t-t'| \geq \tfrac{1}{2}\} \\
&\quad \cup \{(x,x',t,t') : |t-t'| > 2|x-x'|, |t-t'| \geq \tfrac{1}{2}\} \\
&=: \Omega_{6,1} \cup \Omega_{6,2}.
\end{aligned}$$

Now

$$\begin{aligned}
&\int_{\Omega_{6,1}} |W(x,t)|^2 |W(x',t')|^2 \, dx dx' \frac{dtdt'}{|t-t'|^{2+N}} \\
&\lesssim \int_{|t-t'|\gtrsim 1} \int_{|x-x'|\lesssim |t-t'|} |W(x,t)|^2 |W(x',t')|^2 \, dx dx' \frac{dtdt'}{|t-t'|^{N+2}} \\
&\lesssim \int_{|t-t'|\gtrsim 1} \|W(\cdot,t)\|_{\widetilde{r}}^2 \|W(\cdot,t')\|_{\widetilde{r}}^2 \frac{dtdt'}{|t-t'|^{N+2-6(1-\frac{2}{\widetilde{r}})}} \\
&\lesssim \|W\|^4_{L_t^{\widetilde{q}} L_x^{\widetilde{r}}}
\end{aligned}$$

by Young's convolution inequality, for any $\widetilde{q}, \widetilde{r} \in [2,4]$ and upon a sufficiently large choice of N. Finally, using Young's convolution inequality,

$$\begin{aligned}
&\int_{\Omega_{6,2}} |W(x,t)|^2 |W(x',t')|^2 \frac{dx dx'}{|x-x'|^2} \frac{dtdt'}{|t-t'|^N} \\
&\lesssim \int_{|t-t'|\gtrsim 1} \int_{|x-x'|\lesssim |t-t'|} |W(x,t)|^2 |W(x',t')|^2 \frac{dx dx'}{|x-x'|^2} \frac{dtdt'}{|t-t'|^N}
\end{aligned}$$

$$\lesssim \int_{|t-t'|\gtrsim 1} \|W(\cdot,t)\|_{\widetilde{r}}^2 \|W(\cdot,t')\|_{\widetilde{r}}^2 \frac{dtdt'}{|t-t'|^{N-\mu}}$$

for $\widetilde{r} \in (3,4]$, and where $\mu := 3 - (1-\frac{2}{\widetilde{r}})^{-1}$. By making a sufficiently large choice of N, it follows that

$$\int_{\Omega_{6,2}} |W(x,t)|^2 |W(x',t')|^2 \frac{dxdx'}{|x-x'|^2} \frac{dtdt'}{|t-t'|^N} \lesssim \|W\|_{L_t^{\widetilde{q}} L_x^{\widetilde{r}}}^4$$

for any $\widetilde{q} \in [2,4]$ and $\widetilde{r} \in (3,4]$. □

3 Spherical Kinetic Averages when $\beta = 2$ via Fourier Analysis

Results from [2] and [4] can be used off the shelf to prove (1.12) when $\beta = 2$. As described in Sect. 1.3, via a semi-classical limiting argument, such estimates would also follow from (1.7) whenever (q,r) are such that $\beta_{\frac{d}{2}}(q,r) = 2$. In particular when $d = 3$, Theorem 2.1 implies (1.12) for such (q,r) and $r \in [4,6)$. As we shall see below though, the endpoint case $r = 6$ is also valid and can be obtained from [2]. As such, we believe that it is of wider interest to give an exposition of how one may prove (1.12) using the Fourier-analytic approach in [2, 4].

Let us begin by introducing the more general framework of smoothing estimates for ϱ in hyperbolic Sobolev spaces considered first by Bournaveas and Perthame [12] and later Bournaveas and Gutiérrez [11], Bennett et al. [2] and Bez et al. [4]. Suppose $a, b \in [2,\infty)$, and let $s_\pm$ be real numbers satisfying

$$s_+ + s_- = \frac{d}{b} + \frac{1}{a} - \frac{d}{2}. \tag{3.1}$$

Also, set

$$s_-^* := \max\left\{\frac{1}{a} + \frac{d-1}{2b} - \frac{d-1}{2}, \frac{1-d}{4}\right\}.$$

With the above notation, the following holds.

Theorem 3.1 ([2, 4, 11, 12]) *Let $d \geq 2$ and $\frac{1}{a} > \frac{d-1}{2}(\frac{1}{2} - \frac{1}{b})$. If $s_- \geq s_-^*$ then*

$$\|D_+^{s_+} D_-^{s_-} \varrho f\|_{L_t^a L_x^b} \lesssim \|f\|_{L^2_{x,\theta}} \tag{3.2}$$

holds for all $f \in L^2_{x,\theta}$.

In the above estimate, $D_+^{s_+}$ and $D_-^{s_-}$ denote fractional classical derivatives and hyperbolic derivatives, respectively i.e. Fourier multiplier operators with multipliers $(|\xi| + |\tau|)^{s_+}$ and $||\xi| - |\tau||^{s_+}$. The case $a = b = 2$ was first handled in [11, 12]. The extension to general (a, b) was made in [2, 4]; more precisely, the range $s > s_-^*$ was first established in [2] and the more delicate critical case $s = s_-^*$ was proved in [4]. We note that the threshold s_-^* is sharp and this was shown to be the case in [2]. Also, estimates of the form (3.2) are valid in the regime $\frac{1}{a} \leq \frac{d-1}{2}(\frac{1}{2} - \frac{1}{b})$ if and only if $s > s_-^*$ (as shown in [2]), but this regime does not arise when we consider the case of interest in the present article $s_- = s_+ = 0$.

As a consequence of Theorem 3.1, we obtain the following.

Corollary 3.2 *Assume* $\frac{d}{4} = \frac{1}{q} + \frac{d}{r}$.

(1) *If* $d \geq 3$ *and* $r \in [4, \frac{4d}{d-1}]$, *then*

$$\|\varrho f\|_{L_t^{\frac{q}{2}} L_x^{\frac{r}{2}}} \lesssim \|f\|_{L^2_{x,\theta}}. \tag{3.3}$$

(2) *If* $d = 2$ *and* $r \in [4, 6]$, *then* (3.3) *holds.*

Proof of Corollary 3.2 The case $r = 4$ (and $q = \infty$) can be handled easily by using Minkowski's inequality and the Cauchy–Schwarz inequality. For $r > 4$, we set $(a, b, s_-, s_+) = (\frac{q}{2}, \frac{r}{2}, 0, 0)$ and note that

$$s_-^* = \frac{1}{2} - \frac{d+1}{r}.$$

Also, one may check that $\frac{1}{a} > \frac{d-1}{2}(\frac{1}{2} - \frac{1}{b})$ holds. For $d \geq 3$ it is easy to see that the restriction $r \in [4, \frac{4d}{d-1}]$ implies $s_-^* < 0$ and thus the claim in (1) follows from Theorem 3.1. When $d = 2$ a similar argument works but in this case we have $s_-^* \leq 0$ if and only if $r \leq 6$; this explains the restriction on r in (2). □

Here we include the main steps of the proof of Theorem 3.1 in [2] when $s_+ = s_- = 0$ and $d \geq 4$; for such spatial dimensions, a streamlined argument can be given based on use of classical Strichartz estimates for the one-sided wave propagator which, given the theme of the current paper, provides a pleasing sense of coherence.

The key point of the argument is that $\varrho\varrho^*$ is a Fourier multiplier operator given by

$$\widehat{\varrho\varrho^* g}(\xi, \tau) \simeq \frac{1}{|\xi|}\left(1 - \frac{\tau^2}{|\xi|^2}\right)_+^{\frac{d-3}{2}} \widehat{g}(\xi, \tau), \tag{3.4}$$

where $x_+ := \max\{x, 0\}$. To see this, one may first compute

$$\widehat{\varrho f}(\xi, \tau) \simeq \int_{\mathbb{S}^{d-1}} \widehat{f}(\xi, \theta)\delta(\tau + \theta \cdot \xi)\, d\sigma(\theta)$$

where $\widehat{f}(\xi, \theta)$ denotes the Fourier transform of $f(\cdot, \theta)$ on $\mathbb{R}^d$, and δ denotes the Dirac delta distribution supported at the origin. Using this, one may also obtain

$$\widehat{\varrho^* g}(\xi, \theta) \simeq \widehat{g}(\xi, -\xi \cdot \theta),$$

where the Fourier transform on the left-hand side is on $\mathbb{R}^d$ and on the right-hand side is on $\mathbb{R}^{d+1}$. Combining these

$$\begin{aligned}\widehat{\varrho\varrho^* g}(\xi, \tau) &\simeq \int_{\mathbb{S}^{d-1}} \widehat{g}(\xi, -\xi \cdot \theta)\delta(\tau + \theta \cdot \xi)\, d\sigma(\theta) \\ &= \widehat{g}(\xi, \tau) \int_{\mathbb{S}^{d-1}} \delta(\tau + \theta \cdot \xi)\, d\sigma(\theta)\end{aligned}$$

and a direct computation of the above integral over the sphere leads to (3.4).

Now, by duality,

$$\|\varrho f\|_{L_t^a L_x^b} \lesssim \|f\|_{L^2_{x,\theta}} \tag{3.5}$$

is equivalent to

$$\|\varrho\varrho^* g\|_{L_t^a L_x^b} \lesssim \|g\|_{L_t^{a'} L_x^{b'}},$$

but since $\varrho\varrho^*$ is the Fourier multiplier given by (3.4) another application of duality means that

$$\|\mathcal{T} g\|_{L_t^a L_x^b} \lesssim \|g\|_{L^2_{t,x}} \tag{3.6}$$

is also an equivalent estimate to (3.5), where $\mathcal{T}$ is given by

$$\widehat{\mathcal{T} g}(\xi, \tau) = \frac{1}{|\xi|^{1/2}} \left(1 - \frac{\tau^2}{|\xi|^2}\right)_+^{\frac{d-3}{4}} \widehat{g}(\xi, \tau).$$

Furthermore, by use of classical Littlewood–Paley estimates, for $a, b \in [2, \infty)$ satisfying the scaling condition (3.1) with $s_+ = s_- = 0$, (3.6) is equivalent to

$$\|\mathcal{C} g\|_{L_t^a L_x^b} \lesssim \|g\|_{L^2_{t,x}}, \tag{3.7}$$

where $\mathcal{C}$ is given by

$$\widehat{\mathcal{C} g}(\xi, \tau) = \chi(|\xi|) \left(1 - \frac{\tau^2}{|\xi|^2}\right)_+^{\frac{d-3}{4}} \widehat{g}(\xi, \tau),$$

and thus (3.6) follows from $L^2_{t,x} \to L^a_t L^b_x$ bounds for the Fourier multiplier operator with multiplier $\chi(|\xi|)1_{\tau>0}(|\xi|-\tau)_+^{\frac{d-3}{4}}$. The delicate part of the multiplier is near the boundary of the cone $\tau = |\xi|$ and so we make a dyadic decomposition away from the region by writing

$$(|\xi|-\tau)^{\frac{d-3}{4}} = \sum_{k\in\mathbb{Z}} 2^{-\frac{d-3}{4}k}\psi(2^k(|\xi|-\tau))$$

for $\tau \leq |\xi|$, and where $\psi \in C_c^\infty(\mathbb{R})$ is a suitably chosen bump function supported in $[\frac{1}{2}, 2]$. Obtaining appropriately decaying bounds for $k \gg 1$ is the most challenging part (near the boundary of the cone). Taking into account the restriction $|\xi| \sim 1$ from the support of χ, we see that the number of remaining k which contribute to $\chi(|\xi|)1_{\tau>0}(|\xi|-\tau)_+^{\frac{d-3}{4}}$ is $O(1)$ and this part of the multiplier can be easily handled. Thus, it suffices to prove

$$\|C_k g\|_{L^a_t L^b_x} \lesssim 2^{k\alpha^*}\|g\|_{L^2_{t,x}} \tag{3.8}$$

for $k \gg 1$ and an appropriate value of α^*, where

$$\widehat{C_k g}(\xi,\tau) = \chi(|\xi|)1_{\tau>0}\psi(2^k(|\xi|-\tau))\widehat{g}(\xi,\tau).$$

In the regime of interest $\frac{1}{a} > \frac{d-1}{2}(\frac{1}{2}-\frac{1}{b})$, the key estimates are as follows.

Lemma 3.3 *Suppose $a, b \in [2,\infty)$ and $\frac{1}{a} > \frac{d-1}{2}(\frac{1}{2}-\frac{1}{b})$. For $k \gg 1$, (3.8) holds with $\alpha^* = \frac{1}{a} + \frac{d-1}{2b} - \frac{d+1}{4}$.*

Let us show how one can make use of the classical Strichartz estimates for the wave equation to prove Lemma 3.3 when $d \geq 4$. By the Fourier inversion formula and a translation change of variables we have

$$C_k g(x,t) \simeq \int_{\mathbb{R}} \psi(2^k\tau)e^{-it\tau}e^{itD}h_\tau(x)\,d\tau,$$

where $\widehat{h_\tau}(\xi) = \chi(|\xi|)1_{|\xi|>\tau}\widehat{g}(\xi,|\xi|-\tau)$. Thus, if $\frac{1}{a_0} = \frac{d-1}{2}(\frac{1}{2}-\frac{1}{b_0})$ with $a_0 \in [2,\infty]$ and $b_0 \in [2, \frac{2(d-1)}{d-3}]$, then we may apply Minkowski's inequality and the classical Strichartz estimate (1.8), valid for such (a_0, b_0), to get

$$\|C_k g\|_{L^{a_0}_t L^{b_0}_x} \lesssim \int_{\mathbb{R}} \psi(2^k\tau)\|h_\tau\|_{L^2_x}\,d\tau$$

and thus the Cauchy–Schwarz inequality implies

$$\|C_k g\|_{L^{a_0}_t L^{b_0}_x} \lesssim 2^{-\frac{k}{2}}\|g\|_{L^2_{x,t}}.$$

Interpolating this estimate with

$$\|C_k g\|_{L_t^{a_1} L_x^{b_1}} \lesssim \|g\|_{L^2_{x,t}}$$

with $a_1 = b_1 = 2$ (which clearly holds by Plancherel's theorem), we get a proof of Lemma 3.3.

Remark To obtain Lemma 3.3 when $a = 2$ by the above argument, we need the endpoint Strichartz estimate due to Keel–Tao [32]. The above argument almost yields Lemma 3.3 when $d = 3$, only the case $a = 2$ is unattainable in this manner as a result of the failure of the endpoint Strichartz estimate in three space dimensions. For $d = 2$, Strichartz estimates are only available for $a \geq 4$ and so a somewhat different argument is needed to handle the remaining cases in Lemma 3.3; we refer the interested reader to [2, Section 5].

From Lemma 3.3 we obtain

$$\Big\| \sum_{k \gg 1} C_k g \Big\|_{L_t^a L_x^b} \lesssim \sum_{k \gg 1} 2^{-\frac{d-3}{4}k} \|C_k g\|_{L_t^a L_x^b} \lesssim \|g\|_{L^2_{t,x}}$$

as long as $\alpha^* < \frac{d-3}{4}$, which boils down to $\frac{d+1}{r} > \frac{1}{2}$ if we take q, r as in Corollary 3.2 (i.e. $s_* < 0$). This gives a proof of Corollary 3.2(1) when $d \geq 4$.

Remark Notice from the above argument that the restriction $r \in [4, \frac{4d}{d-1}]$ in Corollary 3.2(1) can be seen to arise because of the restriction $a \geq 2$, and in particular there is some slack in the argument at the point where the summation in k is taken. This is related to the fact that we are away from the threshold regularity s_-^*.

When $d = 2$, the estimate corresponding to the endpoint case $(q, r) = (6, 6)$ is

$$\|\varrho f\|_{L^3_{x,t}(\mathbb{R}^2 \times \mathbb{R})} \lesssim \|f\|_{L^2_{x,\theta}(\mathbb{R}^2 \times \mathbb{S}^1)}. \tag{3.9}$$

A Fourier-analytic proof of (3.9) along the above lines is possible but requires the use of the critical case $s_- = s_-^*$; a more delicate argument dealing with this case can be found in [4]. However, estimate (3.9) was already known prior to [4]. For example, one may find it in [12] where it was established by first proving (3.2) when $(d, s_-, s_+) = (2, \frac{1}{4}, \frac{1}{4})$ and then invoking an embedding theorem of Tataru [41] for hyperbolic derivatives. Other proofs of (3.9) are known (see, for example, [23] and [42]) and we include a further proof in the next section.

4 The Sharp Admissible Line

We end the paper with some brief remarks on the estimates (1.11) and (1.12) for exponents on the sharp admissible line

$$\frac{1}{q} = \frac{d-1}{2}\left(\frac{1}{2} - \frac{1}{r}\right)$$

and with $\beta = \beta_{\frac{d}{2}}(q, r)$.

4.1 Wolff's Restricted X-Ray Transform Estimates

In the case $q = r$, the estimate (1.11) becomes

$$\|\varrho^* g\|_{L^d_{x,\theta}(\mathbb{R}^d \times \mathbb{S}^{d-1})} \lesssim \|g\|_{L^{\frac{d+1}{2}}_{x,t}(\mathbb{R}^d \times \mathbb{R})}. \tag{4.1}$$

Via a different parametrisation, estimates of this nature have been shown to be true by Wolff in [43, Theorem A1] if one is willing to accept a certain ε-loss. To be more precise, Wolff proves that for any dimension $d \geq 2$ and any $\varepsilon > 0$ the estimate

$$\left(\int_{\mathbb{S}^{d-1}} \int_{(\theta,1)^\perp} \left| \int_{\ell(y,\theta)} g \right|^d d\lambda_\theta(y)\, d\sigma(\theta)\right)^{\frac{1}{d}} \lesssim_\varepsilon \|(1-\Delta)^{\frac{\varepsilon}{2}} g\|_{L^{\frac{d+1}{2}}} \tag{4.2}$$

holds for all functions supported in some neighbourhood of the origin. Here, λ_θ denotes Lebesgue measure on the hyperplane $(\theta, 1)^\perp$ which is perpendicular to the direction $(\theta, 1)$. Also, $\ell(y, \theta)$ is the straight line given by

$$\ell(y, \theta) := \{y + t(\theta, 1) : t \in \mathbb{R}\} \subset \mathbb{R}^{d+1}$$

for $\theta \in \mathbb{S}^{d-1}$ and $y \in (\theta, 1)^\perp$. More generally, if we write

$$Xg(y, \theta) := \int_{\ell(y,\theta)} g,$$

then Wolff [43, Theorem A1] establishes mixed-norm estimates of the form

$$\|Xg\|_{L^q_\theta L^r_y} \lesssim_\varepsilon \|(1-\Delta)^{\frac{\varepsilon}{2}} g\|_{L^p}$$

for functions supported in some neighbourhood of the origin (the above estimate (4.2) is the special case when $q = r$). These mixed-norm estimates represent substantial progress towards a difficult conjecture raised in the appendix of [43]; if true, the conjecture would imply that all Kakeya sets have maximal Minkowski dimension and hence verify this version of the Kakeya conjecture.

Remark 4.4 To add a little more context, let us permit the θ variable to run over all of $\mathbb{R}^d$ (and use $v \in \mathbb{R}^d$ rather than θ to emphasise the difference). Then the analogous estimate to (4.1) is

$$\left\| \int_{\mathbb{R}} g(x+tv,t)\,dt \right\|_{L^{d+2}_{x,v}(\mathbb{R}^d\times\mathbb{R}^d)} \lesssim \|g\|_{L^{\frac{d+2}{2}}_{x,t}(\mathbb{R}^d\times\mathbb{R})} \tag{4.3}$$

and this is dual to a particular instance of the velocity averaging estimate (1.13). Since v moves over all of $\mathbb{R}^d$, the X-ray transform in (4.3) is taking line integrals over all directions and the estimate (4.3) goes back to Christ [18].

Removing the ε-loss in (4.2) when $d = 2$ is certainly possible because of (3.9). As we have already noted, several proofs of this estimate are known and we sketch a geometric approach here. First expand the L^2 norm on the left-hand side and make an elementary change of variables

$$\|\varrho^* g\|^2_{L^2_{x,\theta}(\mathbb{R}^2\times\mathbb{S}^1)} = \int g(x,t_1)\mathcal{A}_{t_2-t_2}(g(\cdot,t_2))(x)\,dt_2dxdt_1.$$

By Hölder's inequality, Minkowski's inequality and Proposition 2.2, we obtain

$$\|\varrho^* g\|^2_{L^2_{x,\theta}(\mathbb{R}^2\times\mathbb{S}^1)} \lesssim \|g\|_{L^{\frac{3}{2}}_{x,t}(\mathbb{R}^2\times\mathbb{R})} \left\| \int_{\mathbb{R}} \frac{\|g(\cdot,t_2)\|_{L^{\frac{3}{2}}(\mathbb{R}^2)}}{|t_1-t_2|^{\frac{2}{3}}}\,dt_2 \right\|_{L^3_{t_1}(\mathbb{R})}$$

and an application of the Hardy–Littlewood–Sobolev inequality yields the desired estimate. As far as we are aware, removing the ε-loss in Wolff's estimate (4.2) is an open problem in higher dimensions $d \geq 3$.

4.2 *The Point B: A Possible Endpoint Case*

It seems reasonable to conjecture that the point B is the endpoint in the sense that (1.12) holds on $(B, C]$, or possibly $[B, C]$, but fails otherwise. The estimate at B is

$$\|\varrho f\|_{L^{\frac{q}{2}}_t L^{\frac{r}{2}}_x(\mathbb{R}^d\times\mathbb{R})} \lesssim \|f\|_{L^{\beta}_{x,\theta}(\mathbb{R}^d\times\mathbb{S}^{d-1})}, \tag{4.4}$$

where

$$\frac{q}{2} = \frac{d^2+1}{d(d-1)}, \quad \frac{r}{2} = \frac{d^2+1}{(d-1)^2}, \quad \beta = \frac{q}{2}.$$

The dual estimate is

$$\|\varrho^* g\|_{L^{\beta'}_{x,\theta}(\mathbb{R}^d\times\mathbb{S}^{d-1})} \lesssim \|g\|_{L^{\frac{\tilde{q}}{2}}_t L^{\frac{\tilde{r}}{2}}_x(\mathbb{R}^d\times\mathbb{R})}, \tag{4.5}$$

where

$$\beta' = \frac{d^2+1}{d+1}, \quad \frac{\widetilde{q}}{2} = \beta', \quad \frac{\widetilde{r}}{2} = \frac{d^2+1}{2d}.$$

We do not make any contribution here to this endpoint problem. However it is amusing to note that the estimate (1.12) dramatically fails at A when $d = 2$ by taking $f(x, \theta) = 1_E(x)$, where E is a neighbourhood of a Besicovitch–Rado–Kinney set. A Besicovitch–Rado–Kinney set is a null set in the plane which contains a circle of every radius; the existence of such sets was proved independently by Besicovitch and Rado [3] and Kinney [33].

References

1. Bennett, J., Bez, N., Gutiérrez, S., Lee, S.: On the Strichartz estimates for the kinetic transport equation. Commun. Partial Differ. Equations **39**, 1821–1826 (2014)
2. Bennett, J., Bez, N., Gutiérrez, S., Lee, S.: Estimates for the kinetic transport equation in hyperbolic Sobolev spaces. J. Math. Pures Appl. **114**, 1–28 (2018)
3. Besicovitch, A.S., Rado, R.: A plane set of measure zero containing circumferences of every radius. J. Lond. Math. Soc. **43**, 717–719 (1968)
4. Bez, N., Cunanan, J., Lee, S.: Smoothing estimates for the kinetic transport equation at the critical regularity. SIAM J. Math. Anal. **50**, 2280–2294 (2018)
5. Bez, N., Hong, Y., Lee, S., Nakamura, S., Sawano, Y.: On the Strichartz estimates for orthonormal systems of initial data with regularity. Adv. Math. **354**, 106736, 37 pp. (2019)
6. Bez, N., Kinoshita, S., Shiraki, S.: A note on Strichartz estimates for the wave equation with orthonormal initial data. RIMS Kôkyûroku Bessatsu, to appear, arXiv:2306.14547
7. Bez, N., Kinoshita, S., Shiraki, S.: Boundary Strichartz estimates and pointwise convergence for orthonormal systems. Trans. Lond. Math. Soc. **11**, e70002 (2024)
8. Bez, N., Lee, S., Nakamura, S.: Maximal estimates for the Schrödinger equation with orthonormal initial data. Sel. Math. **26**, Article 52 (2020)
9. Bez, N., Lee, S., Nakamura, S.: Strichartz estimates for orthonormal families of initial data and weighted oscillatory integral estimates. Forum Math. Sigma **9**, 52 pp. (2021)
10. Bourgain, J.: Global Solutions of Nonlinear Schrödinger Equations, American Mathematical Society Colloquium Publications, vol. 46 (American Mathematical Society, Providence, 1999)
11. Bournaveas, N., Gutiérrez, S.: On the regularity of averages over spheres for kinetic transport equations in hyperbolic Sobolev spaces. Rev. Mat. Iberoam. **23**, 481–512 (2007)
12. Bournaveas, N., Perthame, B.: Averages over spheres for kinetic transport equations; hyperbolic Sobolev spaces and Strichartz inequalities. J. Math. Pures Appl. **80**, 517–534 (2001)
13. Castella, F., Perthame, B.: Estimations de Strichartz pour les èquations de transport cinétique. C. R. Acad. Sci. Paris Sér. I Math. **332**, 535–540 (1996)
14. Chen, T., Hong, Y., Pavlović, N.: Global well-posedness of the NLS system for infinitely many fermions. Arch. Ration. Mech. Anal. **224**, 91–123 (2017)
15. Chen, T., Hong, Y., Pavlović, N.: On the scattering problem for infinitely many fermions in dimension $d \geq 3$ at positive temperature. Ann. Inst. H. Poincaré Anal. Non Linéaire **35**, 393–416 (2018)
16. Cho, Y., Hwang, G., Hajaiej, H., Ozawa, T.: On the Cauchy problem of fractional Schrödinger equation with Hartree type nonlinearity. Funkcial. Ekvac. **56**, 193–224 (2012)
17. Cho, Y., Hwang, G., Kwon, S., Lee, S.: On the finite time blow-up for mass-critical Hartree equations. Proc. R. Soc. Edinb. Sect. A **145**, 467–479 (2015)

18. Christ, M.: Estimates for the k-plane transform. Indiana Univ. Math. J. **33**, 891–910 (1984)
19. Frank, R., Sabin, J.: The Stein–Tomas inequality in trace ideals, Séminaire Laurent Schwartz – EPD et applications (2015–2016), Exp. No. XV, 12 pp. (2016)
20. Frank, R., Sabin, J.: Restriction theorems for orthonormal functions, Strichartz inequalities, and uniform Sobolev estimates. Amer. J. Math. **139**, 1649–1691 (2017)
21. Frank, R., Lewin, M., Lieb, E.H., Seiringer, R.: Strichartz inequality for orthonormal functions. J. Eur. Math. Soc. **16**, 1507–1526 (2014)
22. Fröhlich, J., Lenzmann, E.: Blowup for nonlinear wave equations describing boson stars. Commun. Pure Appl. Math. **60**, 1691–1705 (2007)
23. Greenleaf, A., Seeger, A.: Fourier integral operators with fold singularities. J. Reine Angew. Math. **455**, 35–56 (1994)
24. Guo, Q., Zhu, S.: Sharp threshold of blowup and scattering for the fractional Hartree equation. J. Differ. Equations **264**, 2802–2832 (2018)
25. Guo, Z., Peng, L.: Endpoint Strichartz estimate for the kinetic transport equation in one dimension. C. R. Math. Acad. Sci. Paris **345**, 253–256 (2007)
26. Hadama, S.: Asymptotic stability of a wide class of steady states for the Hartree equation for random fields, arXiv:2303.02907
27. Hadama, S.: Asymptotic stability of a wide class of stationary solutions for the Hartree and Schrödinger equations for infinitely many particles, arXiv:2308.15929
28. Hadama, S., Hong, Y.: Global well-posedness of the nonlinear Hartree equation for infinitely many particles with singular interaction, arXiv:2404.06730
29. Hadama, S., Yamamoto, T.: Probabilistic Strichartz estimates in Schatten classes and their applications to the Hartree equation, arXiv:2311.02713
30. Hoshiya, A.: Orthonormal Strichartz estimates for Schrödinger operator and their applications to infinitely many particle systems, arXiv:2312.08314
31. Hoshiya, A.: Orthonormal Strichartz estimate for dispersive equations with potentials, arXiv:2401.08675
32. Keel, M., Tao, T.: Endpoint Strichartz estimates. Am. J. Math. **120**, 955–980 (1998)
33. Kinney, J.R.: A thin set of circles. Am. Math. Mon. **75**, 1077–1081 (1968)
34. Lewin, M., Sabin, J.: The Hartree equation for infinitely many particles. II. Dispersion and scattering in 2D. Anal. PDE **7**, 1339–1363 (2014)
35. Lewin, M., Sabin, J.: The Hartree equation for infinitely many particles. I. Well-posedness theory. Commun. Math. Phys. **334**, 117–170 (2015)
36. Lieb, E.H.: An L^p bound for the Riesz and Bessel potentials of orthonormal functions. J. Funct. Anal. **51**, 159–165 (1983)
37. Littman, W.: $L^p - L^q$ estimates for singular integral operators arising from hyperbolic equations. In: Partial Differential Equations (Proc. Sympos. Pure Math., Vol. XXIII, Univ. California, Berkeley, 1971), pp. 479–481. American Mathematical Society, Providence (1973)
38. Mondal, S.S., Swain, J.: Restriction theorem for the Fourier-Hermite transform and solution of the Hermite-Schrödinger equation. Adv. Oper. Theory **7**, Paper No. 44, 18 pp. (2022)
39. Nakamura, S.: The orthonormal Strichartz inequality on torus. Trans. Am. Math. Soc. **373**, 1455–1476 (2020)
40. Ovcharov, E.: Strichartz estimates for the kinetic transport equation, SIAM J. Math. Anal. **43**, 1282–1310 (2011)
41. Tataru, D.: The X^s_θ spaces and unique continuation for solutions to the semilinear wave equation. Commun. Partial Differ. Equations **21**, 841–887 (1996)
42. Wang, H.T.: L^p estimates for the X-ray transform restricted to line complexes of Kirillov type. Trans. Am. Math. Soc. **332**, 793–821 (1992)
43. Wolff, T.: A sharp bilinear cone restriction estimate. Ann. Math. **153**, 661–698 (2001)

Characterization of the Time-Dependent Free Schrödinger Operator by the Galilei Invariance

Hiromichi Nakazato and Tohru Ozawa

Abstract The time-dependent free Schrödinger operator is shown to be characterized as the only linear partial differential operator of the second order that is invariant under the Galilei group in the Euclidean space-time $\mathbb{R} \times \mathbb{R}^n$. The method of proof depends on the analysis of the invariance of polynomials given by the application of the linear partial differential operators to monochromatic plane waves under space rotations and pure Galilei transformations.

Keywords Schrödinger operator · Galilei invariance · Partial differential operator

1 Introduction

In this paper, we consider a general class of linear partial differential operators in the Euclidean space-time $\mathbb{R} \times \mathbb{R}^n$. Among others, we provide a characterization of the time-dependent free Schrödinger operator in terms of the Galilei transformations and local gauge transformations on scalar fields on $\mathbb{R} \times \mathbb{R}^n$. This means in particular that the time-dependent Schrödinger equation is derived from a geometric observation of the Galilei group of motion acting on scalar fields up to local gauge.

In Classical Mechanics, the state of particles is described by the position and velocity in the configuration space $\mathbb{R}^n$. Among others, the motion of the free particles is written by space-time translations, space rotations, and pure Galilei transformations. In other words, the motion of the free particles is characterized by the Galilei group in $\mathbb{R} \times \mathbb{R}^n$.

H. Nakazato
Department of Physics, Waseda University, Shinjuku, Tokyo, Japan
e-mail: hiromici@waseda.jp

T. Ozawa (✉)
Department of Applied Physics, Waseda University, Shinjuku, Tokyo, Japan
e-mail: txozawa@waseda.jp

V. S. Georgiev, T. Ozawa (eds.), *Various Methods for the Analysis of PDEs*,
ICIAM2023 Springer Series 5, https://doi.org/10.1007/978-981-96-9164-7_5

In Quantum Mechanics, the state of the particles is described on the basis of wavefunctions and particular physical quantities such as the position and momentum are given by the corresponding self-adjoint operators through $L^2(\mathbb{R}^n; \mathbb{C})$ scalar product.

From the point of view of representation theory, the group of motion naturally acts on functions based on the notion of pull-back. In this setting, the Galilei group naturally acts on scalar fields on $\mathbb{R} \times \mathbb{R}^n$, and moreover, the notion of local gauge is introduced as a multiplication of functions $\exp(i\theta)$ with a phase function $\theta : \mathbb{R} \times \mathbb{R}^n \to \mathbb{R}$ [1, 2]. Here we follow the terminology of "global" and "local" according to the convention of physics [1, 2]. Namely, the "global" and "local" gauge transformations are given by with factor $e^{i\theta}$ with $\theta \in \mathbb{R}$ and $e^{i\varphi}$ with $\varphi \in C^\infty(\mathbb{R} \times \mathbb{R}^n; \mathbb{R})$. In this paper, we take this point of view and utilize the Galilei group with local gauge acting on scalar fields to study the invariance of partial differential operators. As is clear in the following, the procedure adopted here addressing the possible form of partial differential operators is unique and different from the previous studies where the free Schrödinger equation has been formally derived as the unitary irreducible representation of the Galilei group [3] (see also, e.g., [4–7]).

This paper is organized as follows. In Sect. 2, we introduce the basic notation of the Galilei group in the Euclidean space-time $\mathbb{R} \times \mathbb{R}^n$ and state our main results. In Sect. 3, we prove the main results.

2 Main Results

In this section, we state our main results. For that purpose, we start introducing basic notation. We denote by $x = (t, \boldsymbol{x}) = (x^0, x^1, \dots, x^n)$ a point in $\mathbb{R} \times \mathbb{R}^n$. We also use the notation

$$x = \begin{bmatrix} t \\ \boldsymbol{x} \end{bmatrix} = \begin{bmatrix} x^0 \\ x^1 \\ \vdots \\ x^n \end{bmatrix} \tag{2.1}$$

as a column-vector representation to which $(1 + n) \times (1 + n)$-matrices apply. Let $(\boldsymbol{e}_0, \boldsymbol{e}_1, \dots, \boldsymbol{e}_n)$ be the standard basis of $\mathbb{R} \times \mathbb{R}^n$. Then the column vector representation of a point in $\mathbb{R} \times \mathbb{R}^n$ is given by

$$x = \begin{bmatrix} t \\ \boldsymbol{x} \end{bmatrix} = x^0 \boldsymbol{e}_0 + \sum_{j=1}^{n} x^j \boldsymbol{e}_j. \tag{2.2}$$

We equip the n-dimensional Euclidean space $\mathbb{R}^n$ with the inner product given by

$$\boldsymbol{x} \cdot \boldsymbol{y} = \sum_{j=1}^{n} x^j y^j, \tag{2.3}$$

which is a natural consequence of the orthonormality relations $\boldsymbol{e}_j \cdot \boldsymbol{e}_k = \delta_{jk}$, where $\delta_{jk} = 0$ if $j \neq k$ and $\delta_{jj} = 1$.

Let $\mathrm{O}(n)$ be the orthogonal group defined by

$$\mathrm{O}(n) = \{R \in \mathrm{GL}(n; \mathbb{R});\, {}^tRR = R\,{}^tR = I\}, \tag{2.4}$$

where $\mathrm{GL}(n; \mathbb{R})$ is the general linear group consisting of all invertible $n \times n$ matrices, ${}^tR$ the transposed matrix of R, and I the identity matrix of order n. The action of $R \in \mathrm{O}(n)$ on $\mathbb{R}^n$ is understood as

$$R\boldsymbol{x} = \begin{bmatrix} R^1{}_1 & \cdots & R^1{}_n \\ \vdots & & \vdots \\ R^n{}_1 & \cdots & R^n{}_n \end{bmatrix} \begin{bmatrix} x^1 \\ \vdots \\ x^n \end{bmatrix} = \Big(\sum_{k=1}^{n} R^j{}_k x^k : j \in \{1, \ldots, n\} \Big). \tag{2.5}$$

Here the space for the indices in (2.5) follows the "covariant notation" in physics [1,2]. Space rotations in $\mathbb{R} \times \mathbb{R}^n$ are given by $R \in \mathrm{O}(n)$ as

$$(1 \otimes R)(t, \boldsymbol{x}) = (t, R\boldsymbol{x}), \quad x = (t, \boldsymbol{x}) \in \mathbb{R} \times \mathbb{R}^n. \tag{2.6}$$

Space-time translations in $\mathbb{R} \times \mathbb{R}^n$ are given by $y = (s, \boldsymbol{y}) \in \mathbb{R} \times \mathbb{R}^n$ as

$$T_y x = x - y = (t - s, \boldsymbol{x} - \boldsymbol{y}), \quad x = (t, \boldsymbol{x}) \in \mathbb{R} \times \mathbb{R}^n. \tag{2.7}$$

Pure Galilei transformations in $\mathbb{R} \times \mathbb{R}^n$ are given by $\boldsymbol{v} \in \mathbb{R}^n$ as

$$G_{\boldsymbol{v}} x = (t, \boldsymbol{x} - t\boldsymbol{v}), \quad x = (t, \boldsymbol{x}) \in \mathbb{R} \times \mathbb{R}^n. \tag{2.8}$$

The transformation group generated by $(T_y;\, y \in \mathbb{R} \times \mathbb{R}^n)$, $(1 \otimes R;\, R \in \mathrm{O}(n))$ and $(G_{\boldsymbol{v}};\, \boldsymbol{v} \in \mathbb{R}^n)$ is called the Galilei group in $\mathbb{R} \times \mathbb{R}^n$. The transformation group generated by $(1 \otimes R;\, R \in \mathrm{O}(n))$ and $(G_{\boldsymbol{v}};\, \boldsymbol{v} \in \mathbb{R}^n)$ is called the homogeneous Galilei group in $\mathbb{R} \times \mathbb{R}^n$.

The Galilei group is defined to act on functions $u \in C^\infty(\mathbb{R} \times \mathbb{R}^n; \mathbb{C})$ as the corresponding pull-backs:

$$(T_y^* u)(x) = u(T_y x) = u(x - y) = u(t - s, \boldsymbol{x} - \boldsymbol{y}), \tag{2.9}$$

$$\big((1 \otimes R)^* u\big)(x) = u((1 \otimes R)x) = u(t, R\boldsymbol{x}), \tag{2.10}$$

$$(G_{\boldsymbol{v}}^* u)(x) = u(G_{\boldsymbol{v}} \boldsymbol{x}) = u(t, \boldsymbol{x} - t\boldsymbol{v}). \tag{2.11}$$

We introduce the notion of local gauge as a multiplication of functions of modulus 1 of the form $\exp(i\theta)$ with $\theta \in C^\infty(\mathbb{R} \times \mathbb{R}^n; \mathbb{R})$.

We consider linear partial differential operators in $\mathbb{R} \times \mathbb{R}^n$ of order m of the form

$$L = \sum_{j+|\boldsymbol{\alpha}| \leq m} a_{j\boldsymbol{\alpha}} \partial_t^j \boldsymbol{\partial}^{\boldsymbol{\alpha}}, \tag{2.12}$$

where $\boldsymbol{\alpha} = (\alpha_1, \ldots, \alpha_n) \in \mathbb{Z}_{\geq 0}^n$ is a multi-index with length $|\boldsymbol{\alpha}| = \alpha_1 + \cdots + \alpha_n$, $\boldsymbol{\partial}^{\boldsymbol{\alpha}} = \partial_1^{\alpha_1} \cdots \partial_n^{\alpha_n}$, $\partial_j = \partial/\partial x^j$, $\partial_t = \partial/\partial t$, $a_{j\boldsymbol{\alpha}} \in C(\mathbb{R} \times \mathbb{R}^n; \mathbb{C})$, and $\sum_{j+|\boldsymbol{\alpha}|=m} |a_{j\boldsymbol{\alpha}}| \neq 0$.

We say that L is space-time translation invariant if and only if

$$T_y^* L u = L T_y^* u \tag{2.13}$$

for any $y \in \mathbb{R} \times \mathbb{R}^n$ and any $u \in C^\infty(\mathbb{R} \times \mathbb{R}^n; \mathbb{C})$. We say that L is space-rotation invariant if and only if

$$(1 \otimes R)^* L u = L(1 \otimes R)^* u \tag{2.14}$$

for any $R \in O(n)$ and any $u \in C^\infty(\mathbb{R} \times \mathbb{R}^n; \mathbb{C})$. We say that L is pure Galilei invariant if and only if for any $\boldsymbol{v} \in \mathbb{R}^n$, there exists $\theta_{\boldsymbol{v}} \in C^\infty(\mathbb{R} \times \mathbb{R}^n; \mathbb{R})$ such that

$$e^{i\theta_{\boldsymbol{v}}} G_{\boldsymbol{v}}^* L u = L e^{i\theta_{\boldsymbol{v}}} G_{\boldsymbol{v}}^* u \tag{2.15}$$

for any $u \in C^\infty(\mathbb{R} \times \mathbb{R}^n; \mathbb{C})$. We say that L is Galilei invariant if and only if L is space-time translation invariant, space-rotation invariant, and pure Galilei invariant. We say that L is homogeneous Galilei invariant if and only if L is space-rotation invariant and pure Galilei invariant. Here we remark that an extra degree of freedom θ_v is necessary because the commutativity between L and pure pullback of translations by constant velocity breaks down.

We now state our main results.

Theorem 1 *For any partial differential operator in space-time $\mathbb{R} \times \mathbb{R}^n$ of the second order of the form*

$$L = \sum_{j+|\boldsymbol{\alpha}| \leq 2} a_{j\boldsymbol{\alpha}} \partial_t^j \boldsymbol{\partial}^{\boldsymbol{\alpha}}, \tag{2.16}$$

with $a_{j\boldsymbol{\alpha}} \in C(\mathbb{R} \times \mathbb{R}^n; \mathbb{C})$ and $\sum_{j+|\boldsymbol{\alpha}|=2} |a_{j\boldsymbol{\alpha}}| \neq 0$, the following three statements are equivalent.

(1) L is Galilei invariant. Namely,

(i) For any $y \in \mathbb{R} \times \mathbb{R}^n$ and any $u \in C^\infty(\mathbb{R} \times \mathbb{R}^n; \mathbb{C})$, $T_y^* L u = L T_y^* u$.
(ii) For any $R \in O(n)$ and any $u \in C^\infty(\mathbb{R} \times \mathbb{R}^n; \mathbb{C})$, $(1 \otimes R)^* u = L(1 \otimes R)^* u$.

(iii) For any $\boldsymbol{v} \in \mathbb{R}^n$, there exists $\theta_{\boldsymbol{v}} \in C^\infty(\mathbb{R} \times \mathbb{R}^n; \mathbb{R})$ such that $e^{i\theta_{\boldsymbol{v}}} G_{\boldsymbol{v}}^* L u = L e^{i\theta_{\boldsymbol{v}}} G_{\boldsymbol{v}}^* u$ for any $u \in C^\infty(\mathbb{R} \times \mathbb{R}^n; \mathbb{C})$.

(2) All coefficients of L are constants and L is homogeneous Galilei invariant. Namely,

(i) For any $(j, \boldsymbol{\alpha}) \in \mathbb{Z}_{\geq 0} \times \mathbb{Z}^n_{\geq 0}$ with $j + |\boldsymbol{\alpha}| \leq 2$ and any $x \in \mathbb{R} \times \mathbb{R}^n$, $a_{j\boldsymbol{\alpha}}(x) = a_{j\boldsymbol{\alpha}}(0)$.
(ii) For any $R \in O(n)$ and any $u \in C^\infty(\mathbb{R} \times \mathbb{R}^n; \mathbb{C})$, $(1 \otimes R)^* L u = L(1 \otimes R)^* u$.
(iii) For any $\boldsymbol{v} \in \mathbb{R}^n$, there exists $\theta_{\boldsymbol{v}} \in C^\infty(\mathbb{R} \times \mathbb{R}^n; \mathbb{R})$ such that $e^{i\theta_{\boldsymbol{v}}} G_{\boldsymbol{v}}^* L u = L e^{i\theta_{\boldsymbol{v}}} G_{\boldsymbol{v}}^* u$ for any $u \in C^\infty(\mathbb{R} \times \mathbb{R}^n; \mathbb{C})$.

(3) For any $(j, \boldsymbol{\alpha})$ with $j + |\boldsymbol{\alpha}| \leq 2$, $a_{j\boldsymbol{\alpha}}$ is a complex constant and zero except $a_{\mathbf{0}0}, a_{1\mathbf{0}}, a_{0,2\boldsymbol{e}_1}, \ldots, a_{0,2\boldsymbol{e}_n}$, where $a_{0,2\boldsymbol{e}_1} = \cdots = a_{0,2\boldsymbol{e}_n} \neq 0$ hold. Let $-\alpha \in \mathbb{C}\backslash\{0\}$ be the common value: $-\alpha = a_{0,2\boldsymbol{e}_1} = \cdots = a_{0,2\boldsymbol{e}_n} \neq 0$. Then $ia_{1\mathbf{0}}/\alpha \in \mathbb{R}$, so that if $(\lambda, \beta) \in \mathbb{R} \times \mathbb{C}$ are given by $\lambda = -\frac{ia_{1\mathbf{0}}}{2\alpha}$ and $\beta = a_{0\mathbf{0}}$, L is represented as

$$L = \alpha(2i\lambda\partial_t + \Delta) + \beta, \tag{2.17}$$

where $\Delta = \nabla^2 = \partial_1^2 + \cdots \partial_n^2$ is the Laplacian in $\mathbb{R}^n$. The phase function $\theta_{\boldsymbol{v}}$ in (1)(iii) and (2)(iii)) is given by

$$\theta_{\boldsymbol{v}}(t, \boldsymbol{x}) = \theta_{\boldsymbol{v}}(0, \mathbf{0}) + \lambda \boldsymbol{v} \cdot \boldsymbol{x} - \frac{\lambda}{2} t |\boldsymbol{v}|^2 \quad \text{if} \quad \lambda \neq 0, \tag{2.18}$$

$$\theta_{\boldsymbol{v}}(t, \boldsymbol{x}) = \theta_{\boldsymbol{v}}(t, \mathbf{0}) \quad \text{if} \quad \lambda = 0. \tag{2.19}$$

Remark 1 In the case $\lambda = 0$ in (3), we have $L = \alpha\Delta + \beta$ and $\theta_{\boldsymbol{v}}(t, \boldsymbol{x}) = \theta_{\boldsymbol{v}}(t, \mathbf{0})$ so that $e^{i\theta_{\boldsymbol{v}}}$, $G_{\boldsymbol{v}}^*$, and L are commutative on $C^\infty(\mathbb{R} \times \mathbb{R}^n; \mathbb{C})$.

Remark 2 In the case $\lambda \neq 0$ in (3), the constant β is essentially removable in the sense that

$$\exp\Big(\frac{\beta}{2i\alpha\lambda} t\Big) L \exp\Big(\frac{-\beta}{2i\alpha\lambda} t\Big) u = \alpha(2i\lambda\partial_t + \Delta) u \tag{2.20}$$

for all $u \in C^\infty(\mathbb{R} \times \mathbb{R}^n; \mathbb{C})$. If $\beta \in \mathbb{R}$, the mappings $u \mapsto \exp(\frac{\pm\beta}{2i\alpha\lambda}) u$ are global gauge transformations and L and $\alpha(2i\lambda\partial_t + \Delta)$ are gauge equivalent.

Remark 3 The case $\lambda = 1$ is regarded as the standard choice since the essential factor $2i\partial_t + \Delta$ is the time-dependent free Schrödinger operator in the scale where the Planck constant $\hbar$ and the mass of the free particle are normalized. Moreover, this choice $\lambda = 1$ with another normalization $\theta_{\boldsymbol{v}}(0, \mathbf{0}) = 0$ yields

$$\Big(e^{i\theta_{\boldsymbol{v}}} G_{\boldsymbol{v}}^* u\Big)(t, \boldsymbol{x}) = \exp\big(i(\boldsymbol{v} \cdot \boldsymbol{x} - \frac{t}{2}|\boldsymbol{v}|^2)\big) u(t, \boldsymbol{x} - t\boldsymbol{v}). \tag{2.21}$$

The transformation $u \mapsto e^{i\theta_{\boldsymbol{v}}} G_{\boldsymbol{v}}^* u$ in (2.21) is actually called "the Galilei transformation" in [8–10].

If we assume that $\theta_{\boldsymbol{v}}$ has a specific form as in (2.18), we reformulate Theorem 1 to cover higher order partial differential operators as follows.

Theorem 2 *Let $m \in \mathbb{Z}_{>0}$ and $\lambda \in \mathbb{R}\backslash\{0\}$. For any $\boldsymbol{v} \in \mathbb{R}$, let $\theta_{\boldsymbol{v}} \in C^\infty(\mathbb{R} \times \mathbb{R}^n; \mathbb{R})$ be defined as*

$$\theta_{\boldsymbol{v}}(t, \boldsymbol{x}) = c + \lambda \boldsymbol{v} \cdot \boldsymbol{x} - \frac{\lambda}{2} t |\boldsymbol{v}|^2, \tag{2.22}$$

where $c = \theta_{\boldsymbol{v}}(0, \boldsymbol{0}) \in \mathbb{C}$. Then for any linear partial differential operator in space-time $\mathbb{R} \times \mathbb{R}^n$ of order m of the form

$$L = \sum_{j+|\alpha| \le m} a_{j\alpha} \partial_t^j \boldsymbol{\partial}^\alpha \tag{2.23}$$

with $a_{j\alpha} \in C(\mathbb{R} \times \mathbb{R}^n; \mathbb{C})$ and $\sum_{j+|\alpha|=m} |a_{j\alpha}| \neq 0$, the following three statements are equivalent.

(1) L is Galilei invariant in the following sense.
   - *(i) For any $y \in \mathbb{R} \times \mathbb{R}^n$ and any $u \in C^\infty(\mathbb{R} \times \mathbb{R}^n; \mathbb{C})$, $T_y^* L u = L T_y^* u$.*
   - *(ii) For any $R \in O(n)$ and any $u \in C^\infty(\mathbb{R} \times \mathbb{R}^n; \mathbb{C})$, $(1 \otimes R)^* u = L(1 \otimes R)^* u$.*
   - *(iii) For any $\boldsymbol{v} \in \mathbb{R}^n$ and any $u \in C^\infty(\mathbb{R} \times \mathbb{R}^n; \mathbb{C})$, $e^{i\theta_{\boldsymbol{v}}} G_{\boldsymbol{v}}^* L u = L e^{i\theta_{\boldsymbol{v}}} G_{\boldsymbol{v}}^* u$, where $\theta_{\boldsymbol{v}}$ is defined in (2.22).*

(2) All coefficients of L are constants and L is homogeneous Galilei invariant in the following sense.
   - *(i) For any $(j, \boldsymbol{\alpha}) \in \mathbb{Z}_{\ge 0} \times \mathbb{Z}_{\ge 0}^n$ with $j + |\boldsymbol{\alpha}| \le m$ and any $x \in \mathbb{R} \times \mathbb{R}^n$, $a_{j\alpha}(x) = a_{j\alpha}(0)$.*
   - *(ii) For any $R \in O(n)$ and any $u \in C^\infty(\mathbb{R} \times \mathbb{R}^n; \mathbb{C})$, $(1 \otimes R)^* L u = L(1 \otimes R)^* u$.*
   - *(iii) For any $\boldsymbol{v} \in \mathbb{R}^n$ and any $u \in C^\infty(\mathbb{R} \times \mathbb{R}^n; \mathbb{C})$, $e^{i\theta_{\boldsymbol{v}}} G_{\boldsymbol{v}}^* L u = L e^{i\theta_{\boldsymbol{v}}} G_{\boldsymbol{v}}^* u$, where $\theta_{\boldsymbol{v}}$ is defined in (2.22).*

(3) m is an even integer and L is given by a polynomial in $2i\lambda\partial_t + \Delta$ of order $m/2$. Namely, there exists $(a_j;\ j \in \{0, \ldots, m/2\}) \subset \mathbb{C}$ such that

$$L = \sum_{j=0}^{m/2} a_j (2i\lambda\partial_t + \Delta)^j \quad \text{and} \quad a_{m/2} \neq 0.$$

Theorem 1 shows that the Galilei invariance with local gauge determines the structure of the space of linear partial differential operators of the second order that is exclusively provided with the free Schrödinger operator and the formulation of local gauge as well. Theorem 2 shows that a natural generalization of Theorem 1 is possible for higher-order operators if we fix the local gauge structure. The main results of this paper show that the time-dependent free Schrödinger equation is derived from a simple observation and explicit calculations with the action of the Galilei group upon scalar fields up to local gauge, which is independent of the physical convention of substitution of the energy and momentum by the time and space derivatives, respectively. Our characterization of the time-dependent free Schrödinger equation in terms of the Galilei group with local gauge supports the foundation of Quantum Mechanics on the basis of the invariance naturally arising in Classical Mechanics.

3 Proof of the Main Results

In this section, we prove Theorems 1 and 2. For that purpose, we introduce some notation. For any $\xi = (\tau, \boldsymbol{\xi}) \in \mathbb{R} \times \mathbb{R}^n$, we define $e_\xi \in C^\infty(\mathbb{R} \times \mathbb{R}^n; \mathbb{C})$ by

$$e_\xi(x) = e_\xi(t, \boldsymbol{x}) = \exp(i(\tau t + \boldsymbol{\xi} \cdot \boldsymbol{x})), \quad x = (t, \boldsymbol{x}) \in \mathbb{R} \times \mathbb{R}^n. \tag{3.1}$$

Then we see that $e_\xi(0) = 1$ and for any $(j, \boldsymbol{\alpha}) \in \mathbb{Z}_{\geq 0} \times \mathbb{Z}^n_{\geq 0}$,

$$\partial_t^j \boldsymbol{\partial}^{\boldsymbol{\alpha}} e_\xi = (i\tau)^j (i\boldsymbol{\xi})^{\boldsymbol{\alpha}} e_\xi. \tag{3.2}$$

For any $x, \xi \in \mathbb{R} \times \mathbb{R}^n$, we define

$$p(x, \xi) = \sum_{j+|\boldsymbol{\alpha}| \leq m} a_{j\boldsymbol{\alpha}}(x)(i\tau)^j (i\boldsymbol{\xi})^{\boldsymbol{\alpha}}. \tag{3.3}$$

By (3.1), (3.2), and (3.3), we have

$$Le_\xi(x) = p(x, \xi) e_\xi(x) \tag{3.4}$$

for any $x, \xi \in \mathbb{R} \times \mathbb{R}^n$ and the original operator L is represented as

$$L = p(x, -i\partial_t, -i\nabla), \tag{3.5}$$

where $\nabla = \boldsymbol{\partial} = (\partial_1, \ldots, \partial_n)$. We use the equalities (3.4) and (3.5) in the subsequent argument.

Proof of Theorem 1 (1) $\Rightarrow$ (2): It suffices to prove (1)(i) $\Rightarrow$(2)(i). The translation invariance in (1)(i) with $u = e_\xi$ implies

$$T_y^* L e_\xi = L T_y^* e_\xi \tag{3.6}$$

for any $y = (s, \boldsymbol{y}) \in \mathbb{R} \times \mathbb{R}^n$. The LHS of (3.6) at $x = (t, \boldsymbol{x})$ is evaluated as

$$(T_y^* Le_\xi)(x) = (Le_\xi)(x - y) = p(x - y, \xi)e_\xi(x - y), \tag{3.7}$$

where we have used (3.5), while the RHS of (3.6) at $x = (t, \boldsymbol{x})$ is evaluated as

$$\begin{aligned}(LT_y^* e_\xi)(x) &= \sum_{j+|\boldsymbol{\alpha}|\le 2} a_{j\boldsymbol{\alpha}}(x)(\partial_t^j \boldsymbol{\partial}^{\boldsymbol{\alpha}} T_y^* e_\xi)(x) = \sum_{j+|\boldsymbol{\alpha}|\le 2} a_{j\boldsymbol{\alpha}}(x)(T_y^* \partial_t^j \boldsymbol{\partial}^{\boldsymbol{\alpha}} e_\xi)(x) \\ &= \sum_{j+|\boldsymbol{\alpha}|\le 2} a_{j\boldsymbol{\alpha}}(x)(i\tau)^j (i\boldsymbol{\xi})^{\boldsymbol{\alpha}} e_\xi(x - y) = p(x, \xi)e_\xi(x - y), \end{aligned} \tag{3.8}$$

where we have used (3.2) and (3.3). By (3.6), (3.7), and (3.8), with $y = x$, we have

$$p(0, \xi) = p(x, \xi) \tag{3.9}$$

for any $\xi = (\tau, \boldsymbol{\xi}) \in \mathbb{R} \times \mathbb{R}^n$, since $e_\xi(0) = 1$. It follows from (3.9) that

$$a_{j\boldsymbol{\alpha}}(x) = a_{j\boldsymbol{\alpha}}(0) \tag{3.10}$$

for any $x \in \mathbb{R} \times \mathbb{R}^n$ and any $(j, \boldsymbol{\alpha}) \in \mathbb{Z}_{\ge 0} \times \mathbb{Z}^n_{\ge 0}$, which is precisely the statement in (2)(i).
(2) $\Rightarrow$ (3): By (2)(i), we denote by $p(\xi) = p(0, \xi)$ the following polynomial in $\xi = (\tau, \boldsymbol{\xi})$ of the second order

$$p(\xi) = \sum_{j+|\boldsymbol{\alpha}|\le 2} a_{j\boldsymbol{\alpha}}(i\tau)^j (i\boldsymbol{\xi})^{\boldsymbol{\alpha}} \tag{3.11}$$

with $(a_{j\boldsymbol{\alpha}};\ j + |\boldsymbol{\alpha}| \le 2) \subset \mathbb{C}$. For $j \in \{0, 1, 2\}$, we define

$$p_j(\boldsymbol{\xi}) = \sum_{|\boldsymbol{\alpha}|\le 2-j} a_{j\boldsymbol{\alpha}}(0)(i\boldsymbol{\xi})^{\boldsymbol{\alpha}}, \tag{3.12}$$

so that p is represented as

$$p(\xi) = \sum_{j=0}^{2} p_j(\boldsymbol{\xi})(i\tau)^j = p_0(\boldsymbol{\xi}) + ip_1(\boldsymbol{\xi})\tau - p_2(\boldsymbol{\xi})\tau^2. \tag{3.13}$$

We prove that p_j is invariant under the space rotations:

$$R^* p_j = p_j \tag{3.14}$$

for any $R \in \mathrm{O}(n)$. For that purpose we calculate

$$\begin{aligned}\big((1 \otimes R)^* e_\xi\big)(x) &= e_\xi(t, R\boldsymbol{x}) = \exp(i(\tau t + \boldsymbol{\xi} \cdot R\boldsymbol{x})) \\ &= \exp(i(\tau t + {}^t R\boldsymbol{\xi} \cdot \boldsymbol{x})) = e_{(\tau, {}^t R\boldsymbol{\xi})}(x). \qquad (3.15)\end{aligned}$$

By (3.2) and (3.15), we obtain

$$\partial_t^j \boldsymbol{\partial}^{\boldsymbol{\alpha}} (1 \otimes R)^* e_\xi = (i\tau)^j (i\, {}^t R\boldsymbol{\xi})^{\boldsymbol{\alpha}} e_{(\tau, {}^t R\boldsymbol{\xi})} \qquad (3.16)$$

and therefore

$$L(1 \otimes R)^* e_\xi = \sum_{j+|\boldsymbol{\alpha}| \leq 2} a_{j\boldsymbol{\alpha}} (i\tau)^j (i\, {}^t R\boldsymbol{\xi})^{\boldsymbol{\alpha}} e_{(\tau, {}^t R\boldsymbol{\xi})} = p(\tau, {}^t R\boldsymbol{\xi}) e_{(\tau, {}^t R\boldsymbol{\xi})}, \qquad (3.17)$$

which is equal to $(1 \otimes R)^* L e_\xi$ by (2)(ii), given by

$$(1 \otimes R)^* L e_\xi = p(\xi)(1 \otimes R)^* e_\xi = p(\xi) e_{(\tau, {}^t R\boldsymbol{\xi})}, \qquad (3.18)$$

where we have used (3.4), (3.11), and (3.15). The equality that (3.17) and (3.18) form implies, at $x = 0 \in \mathbb{R} \times \mathbb{R}^n$,

$$\sum_{j=0}^{2} p_j({}^t R\boldsymbol{\xi})(i\tau)^j = p(\tau, {}^t R\boldsymbol{\xi}) = p(\tau, \boldsymbol{\xi}) = \sum_{j=0}^{2} p_j(\boldsymbol{\xi})(i\tau)^j, \qquad (3.19)$$

where we have used (3.13). Since (3.19) holds for any $\xi = (\tau, \boldsymbol{\xi}) \in \mathbb{R} \times \mathbb{R}^n$ and any $R \in \mathrm{O}(n)$, we conclude that

$$({}^t R)^* p_j = p_j \qquad (3.20)$$

holds for any $R \in \mathrm{O}(n)$ and any $j \in \{1, 2, 3\}$. By replacing $R \mapsto {}^t R$, (3.20) is equivalent to (3.14), as required.

Since p_j is invariant under space rotations as in (3.14), it is written as a polynomial in $|\boldsymbol{\xi}|^2$ as

$$p_0(\boldsymbol{\xi}) = b_{00} + b_{01}|\boldsymbol{\xi}|^2, \qquad (3.21)$$

$$p_1(\boldsymbol{\xi}) = b_{10}, \qquad (3.22)$$

$$p_2(\boldsymbol{\xi}) = b_{20} \qquad (3.23)$$

with $b_{00}, b_{01}, b_{10}, b_{20} \in \mathbb{C}$ (see [11–13]), as the orders of p_0, p_1, p_2 are $2, 0, 0$, respectively. By (3.13), (3.21), (3.22), and (3.23), p is represented as

$$p(\tau, \boldsymbol{\xi}) = (b_{00} + b_{01}|\boldsymbol{\xi}|^2) + i b_{10}\tau - b_{20}\tau^2. \qquad (3.24)$$

By the definition of p in (3.11), we see from (3.24) that $b_{00} = a_{\mathbf{0}}$, $b_{01} = a_{0,2\boldsymbol{e}_1} = \cdots = a_{0,2\boldsymbol{e}_n}$, $b_{10} = a_{1\mathbf{0}}$, and $b_{20} = a_{2\mathbf{0}}$. We now define α and β by

$$\alpha = -b_{01} = -a_{0,2\boldsymbol{e}_1} = \cdots = -a_{0,2\boldsymbol{e}_n} \quad \text{and} \quad \beta = a_{0\mathbf{0}}. \tag{3.25}$$

Then, by (3.24) and (3.25), p is given by

$$p(\tau, \boldsymbol{\xi}) = \beta - \alpha|\boldsymbol{\xi}|^2 + ia_{1\mathbf{0}}\tau - a_{2\mathbf{0}}\tau^2. \tag{3.26}$$

By (3.5) and (3.26), L is given by

$$L = p(-i\partial_t, -i\boldsymbol{\nabla}) = \beta + \alpha\Delta + a_{1\mathbf{0}}\partial_t + a_{2\mathbf{0}}\partial_t^2. \tag{3.27}$$

With the representation (3.27), we calculate both sides of the equality in (2)(iii) with $u = e_\xi$ to obtain an explicit formula for L as well as $\theta_{\boldsymbol{v}}$. We carry out differentiation of the function $e^{i\theta_{\boldsymbol{v}}} G_{\boldsymbol{v}}^* e_\xi$ given by

$$\left(e^{i\theta_{\boldsymbol{v}}} G_{\boldsymbol{v}}^* e_\xi\right)(x) = \exp\left(i\theta_{\boldsymbol{v}}(t, \boldsymbol{x}) + i\tau t + i\boldsymbol{\xi} \cdot (\boldsymbol{x} - t\boldsymbol{v})\right) \tag{3.28}$$

with any $\theta_{\boldsymbol{v}} \in C^\infty(\mathbb{R} \times \mathbb{R}^n; \mathbb{R})$ to obtain

$$\boldsymbol{\nabla}\left(e^{i\theta_{\boldsymbol{v}}} G_{\boldsymbol{v}}^* e_\xi\right) = \left(e^{i\theta_{\boldsymbol{v}}} G_{\boldsymbol{v}}^* e_\xi\right)(i\boldsymbol{\nabla}\theta_{\boldsymbol{v}} + i\boldsymbol{\xi}), \tag{3.29}$$

$$\Delta\left(e^{i\theta_{\boldsymbol{v}}} G_{\boldsymbol{v}}^* e_\xi\right) = \left(e^{i\theta_{\boldsymbol{v}}} G_{\boldsymbol{v}}^* e_\xi\right)(i\Delta\theta_{\boldsymbol{v}} - |\boldsymbol{\nabla}\theta_{\boldsymbol{v}}|^2 - 2\boldsymbol{\xi} \cdot \boldsymbol{\nabla}\theta_{\boldsymbol{v}} - |\boldsymbol{\xi}|^2), \tag{3.30}$$

$$\partial_t\left(e^{i\theta_{\boldsymbol{v}}} G_{\boldsymbol{v}}^* e_\xi\right) = \left(e^{i\theta_{\boldsymbol{v}}} G_{\boldsymbol{v}}^* e_\xi\right)(i\partial_t\theta_{\boldsymbol{v}} + i\tau - i\boldsymbol{\xi} \cdot \boldsymbol{v}), \tag{3.31}$$

$$\begin{aligned} \partial_t^2\left(e^{i\theta_{\boldsymbol{v}}} G_{\boldsymbol{v}}^* e_\xi\right) = \left(e^{i\theta_{\boldsymbol{v}}} G_{\boldsymbol{v}}^* e_\xi\right)&(i\partial_t^2\theta_{\boldsymbol{v}} - (\partial_t\theta_{\boldsymbol{v}})^2 - \tau^2 - (\boldsymbol{\xi} \cdot \boldsymbol{v})^2 \\ &- 2\tau\partial_t\theta_{\boldsymbol{v}} + 2\tau\boldsymbol{\xi} \cdot \boldsymbol{v} + 2\partial_t\theta_{\boldsymbol{v}}\boldsymbol{\xi} \cdot \boldsymbol{v}), \end{aligned} \tag{3.32}$$

which together with (3.27) yield

$$\begin{aligned} L\left(e^{i\theta_{\boldsymbol{v}}} G_{\boldsymbol{v}}^* e_\xi\right) = \left(e^{i\theta_{\boldsymbol{v}}} G_{\boldsymbol{v}}^* e_\xi\right)\big(&\beta + \alpha(i\Delta\theta_{\boldsymbol{v}} - |\boldsymbol{\nabla}\theta_{\boldsymbol{v}}|^2 - 2\boldsymbol{\xi} \cdot \boldsymbol{\nabla}\theta_{\boldsymbol{v}} - |\boldsymbol{\xi}|^2) \\ &+ ia_{1\mathbf{0}}(\partial_t\theta_{\boldsymbol{v}} + \tau - \boldsymbol{\xi} \cdot \boldsymbol{v}) \\ &+ a_{2\mathbf{0}}(i\partial_t^2\theta_{\boldsymbol{v}} - (\partial_t\theta_{\boldsymbol{v}})^2 - \tau^2 - (\boldsymbol{\xi} \cdot \boldsymbol{v})^2 \\ &\qquad - 2\tau\partial_t\theta_{\boldsymbol{v}} + 2\tau\boldsymbol{\xi} \cdot \boldsymbol{v} + 2\partial_t\theta_{\boldsymbol{v}}\boldsymbol{\xi} \cdot \boldsymbol{v})\big). \end{aligned} \tag{3.33}$$

By (2)(iii), (3.33) is equal to

$$e^{i\theta_{\boldsymbol{v}}} G_{\boldsymbol{v}}^* L e_\xi = \left(e^{i\theta_{\boldsymbol{v}}} G_{\boldsymbol{v}}^* e_\xi\right)(\beta - \alpha|\xi|^2 + ia_{1\mathbf{0}}\tau - a_{2\mathbf{0}}\tau^2). \tag{3.34}$$

It follows from (3.33) and (3.34) that the equality

$$\begin{aligned}&\alpha(i\triangle\theta_{\boldsymbol{v}} - |\nabla\theta_{\boldsymbol{v}}|^2 - 2\boldsymbol{\xi}\cdot\nabla\theta_{\boldsymbol{v}}) + ia_{10}(\partial_t\theta_{\boldsymbol{v}} - \boldsymbol{\xi}\cdot\boldsymbol{v})\\&\quad + a_{20}(i\partial_t^2\theta_{\boldsymbol{v}} - (\partial_t\theta_{\boldsymbol{v}})^2 - (\boldsymbol{\xi}\cdot\boldsymbol{v})^2 - 2\tau\partial_t\theta_{\boldsymbol{v}} + 2\tau\boldsymbol{\xi}\cdot\boldsymbol{v} + 2\partial_t\theta_{\boldsymbol{v}}\boldsymbol{\xi}\cdot\boldsymbol{v}) = 0 \qquad (3.35)\end{aligned}$$

holds for any $\boldsymbol{v}, \boldsymbol{\xi} \in \mathbb{R}^n$ and any $t \in \mathbb{R}$. By considering the quadratic contribution of $\boldsymbol{\xi}$ in (3.35), we conclude $a_{20} = 0$. Then (3.35) is rewritten as

$$\alpha(i\triangle\theta_{\boldsymbol{v}} - |\nabla\theta_{\boldsymbol{v}}|^2) + ia_{10}\partial_t\theta_{\boldsymbol{v}} - \boldsymbol{\xi}\cdot(2\alpha\nabla\theta_{\boldsymbol{v}} + ia_{10}\boldsymbol{v}) = 0. \qquad (3.36)$$

It follows from (3.36) that $\theta_{\boldsymbol{v}}$ must satisfy

$$2\alpha\nabla\theta_{\boldsymbol{v}} + ia_{10}\boldsymbol{v} = 0, \qquad (3.37)$$

$$\alpha(i\triangle\theta_{\boldsymbol{v}} - |\nabla\theta_{\boldsymbol{v}}|^2) + ia_{10}\partial_t\theta_{\boldsymbol{v}} = 0, \qquad (3.38)$$

since (3.36) holds for any $\boldsymbol{\xi} \in \mathbb{R}^n$. Since $a_{20} = 0$, (3.26) and (3.27) are rewritten as

$$p(\tau, \boldsymbol{\xi}) = \beta - \alpha|\boldsymbol{\xi}|^2 + ia_{10}\tau, \qquad (3.39)$$

$$L = \beta + \alpha\triangle + a_{10}\partial_t, \qquad (3.40)$$

respectively, which in turn ensure that $\alpha \neq 0$ because L is of the second order. This implies that (3.37) is equivalent to

$$\nabla\big(\theta_{\boldsymbol{v}} + \frac{ia_{10}}{2\alpha}\boldsymbol{v}\cdot\boldsymbol{x}\big) = 0, \qquad (3.41)$$

which immediately proves that $\theta_{\boldsymbol{v}}$ is given by

$$\theta_{\boldsymbol{v}}(t, \boldsymbol{x}) = \theta_{\boldsymbol{v}}(t, \boldsymbol{0}) - \frac{ia_{10}}{2\alpha}\boldsymbol{v}\cdot\boldsymbol{x}. \qquad (3.42)$$

The assumption that $\theta_{\boldsymbol{v}}$ is real-valued implies that the coefficient of $\boldsymbol{v}\cdot\boldsymbol{x}$ on the RHS of (3.42) must be real and we introduce

$$\lambda = -\frac{ia_{10}}{2\alpha} \in \mathbb{R}. \qquad (3.43)$$

This shows that

$$\theta_{\boldsymbol{v}}(t, \boldsymbol{x}) = \theta_{\boldsymbol{v}}(t, \boldsymbol{0}) + \lambda\boldsymbol{v}\cdot\boldsymbol{x}, \qquad (3.44)$$

$$\nabla\theta_{\boldsymbol{v}}(t, \boldsymbol{x}) = \lambda\boldsymbol{v}, \qquad (3.45)$$

$$\triangle\theta_{\boldsymbol{v}}(t, \boldsymbol{x}) = 0. \qquad (3.46)$$

Then, Eq. (3.38) takes the form

$$|\nabla\theta_{v}|^2 + 2\lambda\partial_t\theta_{v} = \lambda^2|v|^2 + 2\lambda\partial_t\theta_{v}(t, \mathbf{0}) = 0, \tag{3.47}$$

which shows that $\theta_{v}(t, \mathbf{0})$ is given by

$$\theta_{v}(t, \mathbf{0}) = \theta_{v}(0, \mathbf{0}) - \frac{\lambda}{2}t|v|^2 \quad \text{if} \quad \lambda \neq 0. \tag{3.48}$$

This implies that θ_{v} is given by

$$\theta_{v}(t, x) = \theta_{v}(0, \mathbf{0}) + \lambda v \cdot x - \frac{\lambda}{2}t|v|^2 \quad \text{if} \quad \lambda \neq 0, \tag{3.49}$$

$$\theta_{v}(t, x) = \theta_{v}(t, \mathbf{0}) \quad \text{if} \quad \lambda = 0. \tag{3.50}$$

Moreover, p and L are given by

$$p(\tau, x) = \beta + \alpha(-2\lambda\tau - |\xi|^2), \tag{3.51}$$

$$L = \beta + \alpha(2i\lambda\partial_t + \triangle), \tag{3.52}$$

respectively. This completes the proof of (2) $\Rightarrow$ (3).
(3) $\Rightarrow$ (1): Since (2) $\Rightarrow$ (1), it suffices to prove (3) $\Rightarrow$ (2). By the commutativity between $\triangle$ and R^*, the statement (2)(ii) follows immediately. Therefore, we only have to show (2)(iii). Let $\theta \in C^\infty(\mathbb{R} \times \mathbb{R}^n; \mathbb{R})$ and $u \in C^\infty(\mathbb{R} \times \mathbb{R}^n; \mathbb{C})$ and differentiate the function $e^{i\theta}G_{v}^{*}u$ given by

$$\left(e^{i\theta}G_{v}^{*}u\right)(t, x) = e^{i\theta(t,x)}u(t, x - tv) \tag{3.53}$$

to obtain

$$\nabla\left(e^{i\theta}G_{v}^{*}u\right) = e^{i\theta}(i\nabla\theta G_{v}^{*}u + G_{v}^{*}\nabla u), \tag{3.54}$$

$$\triangle\left(e^{i\theta}G_{v}^{*}u\right) = e^{i\theta}\left((i\triangle\theta - |\nabla\theta|^2)G_{v}^{*}u + 2i\nabla\theta \cdot G_{v}^{*}\nabla u + G_{v}^{*}\triangle u\right), \tag{3.55}$$

$$\partial_t\left(e^{i\theta}G_{v}^{*}u\right) = (i\partial_t\theta G_{v}^{*}u + G_{v}^{*}\partial_t u - v \cdot G_{v}^{*}\nabla u) \tag{3.56}$$

so that the action of $L = \alpha(2i\lambda\partial_t + \triangle) + \beta$ on $e^{i\theta}G_{v}^{*}u$ is given by

$$L\left(e^{i\theta}G_{v}^{*}u\right) = e^{i\theta}G_{v}^{*}Lu + \alpha e^{i\theta}(-2\lambda\partial_t\theta + i\triangle\theta - |\nabla\theta|^2)G_{v}^{*}u + 2i\alpha e^{i\theta}(\nabla\theta - \lambda v) \cdot G_{v}^{*}\nabla u. \tag{3.57}$$

For $\lambda \in \mathbb{R}\backslash\{0\}$, we define θ as $\theta(t, \boldsymbol{x}) = \theta(0, \mathbf{0}) + \lambda \boldsymbol{v} \cdot \boldsymbol{x} - \frac{\lambda}{2} t |\boldsymbol{v}|^2$ to obtain

$$- 2\lambda\partial_t\theta + i\Delta\theta - |\nabla\theta|^2 = -2\lambda(-\frac{\lambda}{2}|\boldsymbol{v}|^2) - |\lambda\boldsymbol{v}|^2 = 0, \tag{3.58}$$

$$\nabla\theta - \lambda\boldsymbol{v} = \lambda\boldsymbol{v} - \lambda\boldsymbol{v} = 0. \tag{3.59}$$

For $\lambda = 0$, we define θ as $\theta(t, \boldsymbol{x}) = \theta(t, \mathbf{0})$ to obtain

$$- 2\lambda\partial_t\theta + i\Delta\theta - |\nabla\theta|^2 = i\Delta\theta - |\nabla\theta|^2 = 0, \tag{3.60}$$

$$\nabla\theta - \lambda\boldsymbol{v} = -\lambda\boldsymbol{v} = 0. \tag{3.61}$$

In both cases, we have $L(e^{i\theta} G_{\boldsymbol{v}}^* u) = e^{i\theta} G_{\boldsymbol{v}}^* L u$, as required. □

Proof of Theorem 2 It suffices to prove that (2) ⇒ (3) since other statements are proved in the same way as in the proof of Theorem 1. Let $\theta_{\boldsymbol{v}}$ be as in (2.22). Then $e^{i\theta_{\boldsymbol{v}}} G_{\boldsymbol{v}}^* e_\xi$ is given by

$$\begin{aligned}
\big(e^{i\theta_{\boldsymbol{v}}} G_{\boldsymbol{v}}^* e_\xi\big)(x) &= \exp\big(i(c + \lambda\boldsymbol{v} \cdot \boldsymbol{x} - \frac{\lambda}{2} t|\boldsymbol{v}|^2)\big) \exp\big(i(\tau t + \boldsymbol{\xi} \cdot (\boldsymbol{x} - t\boldsymbol{v}))\big) \\
&= e^{ic} \exp\big(i((\tau - \boldsymbol{\xi} \cdot \boldsymbol{v} - \frac{\lambda}{2}|\boldsymbol{v}|^2)t + (\boldsymbol{\xi} + \lambda\boldsymbol{v}) \cdot \boldsymbol{x}\big) \\
&= e^{ic} e_{(\tau - \boldsymbol{\xi}\cdot\boldsymbol{v} - \frac{\lambda}{2}|\boldsymbol{v}|^2,\ \boldsymbol{\xi} + \lambda\boldsymbol{v})}(x).
\end{aligned} \tag{3.62}$$

This leads to

$$L e^{i\theta_{\boldsymbol{v}}} G_{\boldsymbol{v}}^* e_\xi = e^{ic} \sum_{j+|\boldsymbol{\alpha}|\le m} a_{j\boldsymbol{\alpha}} \big(i(\tau - \boldsymbol{\xi} \cdot \boldsymbol{v} - \frac{\lambda}{2}|\boldsymbol{v}|^2)\big)^j (\boldsymbol{\xi} + \lambda\boldsymbol{v})^{\boldsymbol{\alpha}} e_{(\tau - \boldsymbol{\xi}\cdot\boldsymbol{v} - \frac{\lambda}{2}|\boldsymbol{v}|^2,\ \boldsymbol{\xi} + \lambda\boldsymbol{v})}, \tag{3.63}$$

where $a_{j\boldsymbol{\alpha}} \in \mathbb{C}$ by (2)(i). By (2)(iii), the LHS of (3.63) is equal to

$$e^{i\theta_{\boldsymbol{v}}} G_{\boldsymbol{v}}^* L e_\xi = p(\xi) e^{i\theta_{\boldsymbol{v}}} G_{\boldsymbol{v}}^* e_\xi = e^{ic} p(\xi) e_{(\tau - \boldsymbol{\xi}\cdot\boldsymbol{v} - \frac{\lambda}{2}|\boldsymbol{v}|^2,\ \boldsymbol{\xi} + \lambda\boldsymbol{v})}. \tag{3.64}$$

The resulting equality between (3.63) and (3.64) evaluated at $x = 0 \in \mathbb{R} \times \mathbb{R}^n$ becomes

$$\begin{aligned}
p(\xi) = \sum_{j=0}^{m} p_j(\boldsymbol{\xi})(i\tau)^j &= p(\tau - \boldsymbol{\xi} \cdot \boldsymbol{v} - \frac{\lambda}{2}|\boldsymbol{v}|^2,\ \boldsymbol{\xi} + \lambda\boldsymbol{v}) \\
&= \sum_{j=0}^{m} p_j(\boldsymbol{\xi} + \lambda\boldsymbol{v})\big(i(\tau - \boldsymbol{\xi} \cdot \boldsymbol{v} - \frac{\lambda}{2}|\boldsymbol{v}|^2)\big)^j
\end{aligned} \tag{3.65}$$

for any $\xi = (\tau, \boldsymbol{\xi}) \in \mathbb{R} \times \mathbb{R}^n$ and any $\boldsymbol{v} \in \mathbb{R}^n$. We already know from (2)(ii) that p_j is invariant under space rotations and represented as a polynomial of $|\boldsymbol{\xi}|^2$ of the form

$$p_j(\boldsymbol{\xi}) = \sum_{2k \leq m-j} b_{jk} |\boldsymbol{\xi}|^{2k} \tag{3.66}$$

with $b_{jk} \in \mathbb{C}$. By (3.65) and (3.66), we have the equality

$$\sum_{j=0}^{m} \sum_{2k \leq m-j} b_{jk} |\boldsymbol{\xi}|^{2k} (i\tau)^j = \sum_{j=0}^{m} \sum_{2k \leq m-j} b_{jk} |\boldsymbol{\xi} + \lambda \boldsymbol{v}|^{2k} \big(i(\tau - \boldsymbol{\xi} \cdot \boldsymbol{v} - \frac{\lambda}{2}|\boldsymbol{v}|^2)\big)^j \tag{3.67}$$

for any $\xi = (\tau, \boldsymbol{\xi}) \in \mathbb{R} \times \mathbb{R}^n$ and $\boldsymbol{v} \in \mathbb{R}^n$. If we choose $\boldsymbol{v} = -\frac{1}{\lambda}\boldsymbol{\xi}$, we have $\boldsymbol{\xi} + \lambda \boldsymbol{v} = 0$ and $-\boldsymbol{\xi} \cdot \boldsymbol{v} - \frac{\lambda}{2}|\boldsymbol{v}|^2 = \frac{1}{2\lambda}|\boldsymbol{\xi}|^2$ and (3.67) becomes

$$\sum_{j=0}^{m} \sum_{2k \leq m-j} b_{jk} |\boldsymbol{\xi}|^{2k} (i\tau)^j = \sum_{j=0}^{m} b_{j0} \big(i(\tau + \frac{1}{2\lambda}|\boldsymbol{\xi}|^2)\big)^j. \tag{3.68}$$

Here we recall that in the context of Theorem 2, $\lambda \neq 0$. Since the order of the polynomial p is m, the order of the RHS of (3.68) must satisfy $2j \leq m$, which means that $b_{j0} = 0$ if $j > m/2$, implying that m is even and p is given by the polynomial of $2\lambda\tau + |\boldsymbol{\xi}|^2$ of order $m/2$. Since L is written as $L = p(-i\partial_t, -i\nabla)$, the statement (3) holds. □

Acknowledgments We wish to thank the referees for important remarks and suggestions.

H.N. is partially supported by JSPS KAKENHI Grant No. 23K03268. T.O. is partially supported by JSPS KAKENHI Grant No. 24H00024.

Declarations **H. Nakazato**: Writing-original draft (support); formal analysis (support); writing-review and editing (equal). **T. Ozawa**: Conceptualization (lead); writing-original draft (lead); formal analysis (lead); writing-review and editing (equal).

Conflict of Interest The authors declare that they have no known competing financial interests or personal relationships that could have appeared to influence the work reported in this paper.

Data Availability Data sharing is not applicable to this article as no new data were created or analyzed in this study.

References

1. Bargmann, V.: On unitary ray representations of continuous groups. Ann. Math. **59**, 1–46 (1954)
2. Barut, A.O.: Conformal group →Schrödinger group →dynamical group – the maximal kinematical group of the massive Schrödinger particle. Helv. Phys. Acta **46**, 493–503 (1973)
3. Hamermesh, M.: Galilei invariance and the Schrödinger equation. Ann. Phys. **9**, 518–521 (1960)
4. Levy-Leblond, J.-M.: Galilei group and nonrelativistic quantum mechanics. J. Math. Phys. **4**, 776–788 (1963)
5. Roncadelli, M.: A Simplified approach to the Schrödinger equation. Nuovo Cimento **50A**, 509–514 (1979)
6. Musielak, Z.E., Fry, J.L.: Physical theories in Galilean space-time and the origin of Schrödinger-like equations. Ann. Phys. **324**, 296–308 (2009)
7. Musielak, Z.E., Fry, J.L.: General dynamical equations for free particles and their Galilean invariance. Int. J. Theor. Phys. **48**, 1194–1202 (2009)
8. Carles, R.: Semi-classical Analysis for Nonlinear Schrödinger Equations, 2nd edn. World Scientific, Singapore (2021)
9. Ginibre, J.: Introduction aux Équations de Schrödinger non Linéaires, Paris Onze Édition L161 (1998)
10. Linares, F., Ponce, G.: Introduction to Nonlinear Dispersive Equations, 2nd edn., Universitext. Springer, New York (2014)
11. Nakazato, H., Ozawa, T.: Characterization of the D'Alembertian by the Poincaré invarivance. In: Proceedings of the conference "Critical Phenomena in Nonlinear Partial Differential Equations, Harmonic Analysis, and Functional Inequalities". Springer (2025, in press)
12. Nomura, T.: Spherical Harmonics and Group Representations. Nippon Hyoron Publ. (2018) (in Japanese)
13. Shimakura, N.: Elliptic Partial Differential Operators. Kinokuniya Publ. (1978) (in Japanese)

GPSR Compliance

The European Union's (EU) General Product Safety Regulation (GPSR) is a set of rules that requires consumer products to be safe and our obligations to ensure this.

If you have any concerns about our products, you can contact us on ProductSafety@springernature.com

In case Publisher is established outside the EU, the EU authorized representative is:

Springer Nature Customer Service Center GmbH
Europaplatz 3
69115 Heidelberg, Germany

Batch number: 10725798

Printed by Printforce, the Netherlands